COURTESANS' QUARTER

Premchand was the pen name adopted by the Hindi writer Dhanpat Rai who was born on 31 July 1880 in the village Lamahi near Benares. His father, Munshi Ajaib Lal, was a clerk in the postal department. Premchand's early education was in a *madrassa* under a *maulvi*, where he learnt Urdu. Premchand was only eight years old when his mother died. He was married at the early age of fifteen but that marriage did not work out and he later married a second time.

After his father's death he was forced to take on the family responsibilities. He gave up his studies and got a government job as a schoolteacher. While working, he studied privately and passed his Intermediate and BA examinations. After a series of promotions, he became Deputy Inspectors of Schools. His first story appeared in the magazine *Zamana* published from Kanpur. *Soz-e-Vatan*, a collection of patriotic stories, written under the pseudonym of Nawab Rai was banned by the British government. Therefore, from 1910 he started writing under the pen name of 'Premchand'. In response to Mahatma Gandhi's call of non-cooperation with the British, he quit his job. He worked to generate patriotism and nationalistic sentiments in the general populace.

Premchand was the first Hindi author to introduce realism in his writings. Besides being a great novelist, Premchand was also a social reformer and thinker. His greatness lies in the fact that his writings embody social purpose and social criticism rather than mere entertainment.

Premchand was a prolific writer. He has left behind a dozen novels and nearly 250 short stories. Among his best known novels are: *Seva Sadan*, *Rangmanch*, *Gaban*, *Nirmala* and *Gaudaan*. Much of Premchand's best work is to be found among his 250 or so short stories, collected in Hindi under the title *Manasarovar*. Three of his novels have been made into films. His magnum opus is considered to be *Gaudaan* (now published in English as *The Gift of a Cow* in UNESCO's Asian Literature Series).

Premchand chaired the first all-India conference of the Indian Progressive Writers Association in April 1936 at Lucknow. The continual struggle that he had had to make for a living had however worn him down, and Premchand succumbed to his gastric ulcer, dropsy and cirrhosis of the liver on 8 October 1936.

CLASSICS FROM SOUTH ASIA AND THE NEAR EAST

MUNSHI PREMCHAND

Courtesans' Quarter

A Translation of Bazaar-e-Husn

Translated with Notes by
AMINA AZFAR

Foreword by
RALPH RUSSELL

Introduction by
M. H. Askari

OXFORD
UNIVERSITY PRESS

OXFORD

UNIVERSITY PRESS

Great Clarendon Street, Oxford OX2 6DP

Oxford University Press is a department of the University of Oxford.
It furthers the University's objective of excellence in research, scholarship,
and education by publishing worldwide in

Oxford New York

Auckland Bangkok Buenos Aires Cape Town Chennai
Dar es Salaam Delhi Hong Kong Istanbul Karachi Kolkata
Kuala Lumpur Madrid Melbourne Mexico City Mumbai Nairobi
São Paulo Shanghai Taipei Tokyo Toronto

Oxford is a registered trade mark of Oxford University Press
in the UK and in certain other countries

ISBN 0 19 597710 6

Typeset in Times
Printed in Pakistan by
Kagzi Printers, Karachi.
Published by
Ameena Saiyid, Oxford University Press
Plot No. 38, Sector 15, Korangi Industrial Area, PO Box 8214
Karachi-74900, Pakistan.

Contents

Foreword

The book before you is an early work, first published in 1917, of Premchand, the greatest Urdu and Hindi novelist and short story writer of the earlier half of the twentieth century. It also happens that it was the first work of Urdu prose literature that I encountered in 1946 when I began my university course in Urdu. I at once thought, 'This should be translated into English', and now, more than fifty years later, it has been.

It is a fascinating book, and if English readers feel, as they well may, that there is too much melodrama in it, let me assure them that it is not in the least unrealistic. Leaving aside the question of how far this is still the case, at the time when Premchand wrote it, the Indian social scene was full of melodrama. A continuing theme throughout the book is that of people motivated by human decency and a desire to show a proper respect for their fellow men and women struggling against the implacable tyranny of cruel and immensely powerful social conventions and frequently finding that they lack the courage to continue the fight. This is a novel about men, respectable women, and courtesans; and their interrelationships are vividly and realistically portrayed.

The phenomenon of the courtesan is no longer familiar to the contemporary English-speaking world. Like the hetaira of ancient Athens, she was the product of a society in which respectable women—the great majority of all women—took no part in public life and even in their private lives could form no friendly relationships with men, other than their fathers, husbands, sons, and other close kin. It was a society like that of ancient Athens, of which Pericles, its greatest statesman, in his speech in praise of the city of his birth, said 'The greatest glory of a woman is to be least talked about by men, whether they are praising or criticizing.' Intelligent, cultured men needed the company of intelligent, cultured women, and this is what the hetaira was. Pericles' own liaison with one, Aspasia, was well known, and he suffered no censure for it. The courtesan also sold her sexual services and so could fairly be called a prostitute, although the most

accomplished of them could be very choosy about who they offered
these services to. 'Prostitute', and its Urdu equivalent, are pejorative
words, while 'courtesan' is not unambiguously so. The accomplished
courtesan had to be not only sexually attractive but also extremely
well educated, very commonly an accomplished singer, and in many
cases a good poet, and there were courtesans who were not available
for sex, but attracted a clientele by their other accomplishments.
Premchand seems to suggest that Saman, the central figure in this
novel, was one such. If men who resorted to courtesans were regarded
less highly than those who (as far as anyone knew) did not, society's
censure of them was more formal than real. Predictably, the courtesans
themselves were not accorded any such tolerance. They were immoral
women and an immoral woman was an immoral woman, and it was
only in the practice of her trade that she could appear in the company
of respectable people, people who nevertheless thought it not only
permissible but obligatory to engage them to provide entertainment on
grand occasions like weddings. And yet, in the eyes of society, once a
courtesan was always a courtesan, and a quick and lasting repentance,
such as Saman's life exemplifies, made no difference. She was
stigmatised for life.

So theme-wise, there is no lack of interest for English-speaking
readers. To which I must add that there are features in Premchand's
narrative which are likely to jolt their preconceptions of what a novel
should be. For example, the dialogue is often set out in a way which
English readers are accustomed to only in the written text of plays, so
that where they would have expected, for example, a conversation
between Gujanand and Oma Nath to be written:

'Are you coming from Benaras?' Gujanand asked.
'No', said Oma Nath. 'I am coming from a village.'

they find instead:

Gujanand: Are you coming from Benaras?
Oma Nath: No, I am coming from a village.

When I first encountered this feature of Urdu narrative I found it most
disconcerting. I hope that present-day readers will find it much less so
and will reflect that it is a perfectly reasonable way of reporting a
conversation.

Another feature which has to be accepted is Premchand's not infrequent explicit moralizing over the words and actions of his characters whereas English readers are more used to narrative in which the moral is left implicit in the description of the incident being narrated.

But acceptance of what would once have been unacceptable to novel-readers is, happily, now much more common than it once was. Years ago, the American critics Scholes and Kellogg performed a valuable service in their book, *The Nature of Narrative* (1966), when they said that they aimed to 'put the novel in its place,' and attacked the attitude which regarded any narrative which did not fit into the then generally accepted definition of a novel as a novel manqué. So be on guard against any remnants of this attitude and enjoy this excellent narrative for what it is.

RALPH RUSSELL

Introduction

Although Urdu has a history going back to almost a thousand years, it was not until about the latter part of the nineteenth century that the Urdu novel, as we know it, made its debut, with the writings of Maulvi Nazir Ahmad (b. 1836). However, critics regarded him as something of a moralizer, his novels were his message of reform and his characters were the prototype which were meant to carry his message to the reader. After him, Pandit Ratan Nath Sarshar (b. *circa* 1845) conceived of the novel as a fantasy or epic and his major work *Fasana-i-Azad* (originally serialized in the journal *Avadh Punch*) was quite transparently modelled after *Don Quixote*. The novels of Abdul Halim Sharar (b. 1860), a somewhat junior contemporary of Sarshar were in reality, historical fantasies, conceived around what he believed to be the early glorious period of Islam. One of his novels, *Malik-ul-Aziz Varjana*, was in fact intended to be a rejoinder to Walter Scott's *Talisman*, which Sarshar regarded as an insult to Islam. Towards the end of the nineteenth century and the beginning of the twentieth century, there was a small number of Western educated writers of Urdu, who wrote novels modelled after the novels of English, French and Russian languages. A substantial proportion of their work consisted of adaptations, if not actual translations, of what was being published in the West. One of them, Sajjad Hyder Yaldram borrowed freely from the Turkish language. However, they all enriched Urdu language and literature very significantly.

An outstanding exception to the trend followed by the novelists of this period is to be found in the work of Munshi Premchand (b. 1880) who was altogether an indigenous writer mirroring the simple, underhand life of the people in the semi-urban environment to which he himself belonged.

It was a refreshing departure from the trend set by the writers who preceded him and were basically the product of the urban middle-class society. Premchand's real name was Dhanpat Rai but he preferred to write under his pen name. He deserves to be recognized as one of the most eminent personalities in his generation. His work was published both in Urdu and Hindi. He was the author of a large number of short stories and novels. However, there is a virtual consensus that his novel *Bazaar-e-Husn*, presented here in its English translation under the title *Courtesans' Quarter*, is among his best work. It was first published in 1917.

Although something of a visionary and idealist, Premchand's work was down to earth. There was nothing contrived, unreal, or alien about his short stories and novels. Born and brought up in a village near Benaras (since renamed Varanasi) he hardly had the benefit of higher education in his early life, studying Persian for eight years under a maulvi in his village. Later he moved to Benaras and matriculated as a private candidate and began his working life as a schoolteacher in a modest school. Ultimately, in 1919 he managed to earn his BA degree and rose to the position of deputy inspector of schools. For some time he was active in politics, a pacifist and nationalist under the influence of Mahatma Gandhi. However, he had begun to gain reputation as a writer and was able to devote almost all his time to writing. He published several collections of short stories and novels of considerable merit, and published both in Urdu and Hindi. A great deal of his work has survived the ravages of time and he continues to be regarded as a major writer. The renowned scholar and critic of Urdu, Dr Syed Abdullah, has classified Sir Syed Ahmad Khan, Mohammad Iqbal and Premchand as the three dominating personalities of their respective times, influencing the evolution of modern Urdu literature.

Premchand had developed an acute awareness of the plight of his people at an early stage in his literary career, and without sounding didactic, drew the attention of his readers to their social, economic and cultural backwardness. He also felt strongly that any creative writing worth its name had to be imbued with a set of moral and social values. Premchand's reputation as a writer of national eminence was further consolidated, when he was called upon to chair the inauguration session of the Progressive Writers Association (PWA) in Lucknow in 1936. The session was to be attended by scores of writers from many parts of the country, expressing themselves in diverse languages. Premchand still led a somewhat reclusive life in Benaras at the time. The late Sajjad Zaheer, who himself enjoyed a wide reputation as a writer and poet of Urdu, and was nationally known for his close association with the Communist Party of India, was one of the principal organizers of the Lucknow conference. In his memoirs, he recalls that there was virtual consensus on Premchand being invited to chair the

PWA among writers from almost all over India—Bengal, Bombay, Hyderabad, Ahmadabad, and Lahore. Sajjad Zaheer says that it was with some reluctance that Premchand accepted the invitation to chair the PWA inaugural, insisting that he did not merit the honour and that some one else like Dr Zakir Husain or Jawaharlal Nehru should instead be invited.

In his presidential address Premchand stressed that no work of art or literature could be of any worth if it was not rooted in a commitment to the emancipation of the common people and to their liberation from exploitation by the society. He called upon the writers to form 'the vanguard of a crusade against capitalism, militarism, colonialism and not be content merely with the creation of "beautiful" works of art.' Incredible that Premchand should have said all this in 1936; he could as well be addressing writers in today's contemporary milieu.

The Urdu critic and literary historian, Dr M. Sadiq regards *Bazaar-e-Husn* as 'most satisfactory' of all of Premchand's novels. He likens the central theme of the novel to Hardy's *Tess of the D'Urbervilles*—'the wreck of a beautiful life by a single false step in a self-righteous world.'

However, *Courtesans' Quarter* is not a simple, naive tale of an unfortunate village belle. It presents a remarkable panorama of the social and cultural complexities of life even in a relatively uncomplicated setting of a village. The characters are not mere prototypes. They are real flesh and blood. They have their normal strengths and weaknesses, their small joys and sorrows.

The story of the novel is built around the household of Krishan Chandr, a police inspector, who during most of his career, leads an honest life, staying away from the corrupting environment of his position, but is at the same time deeply concerned about the future of his two daughters because of the time-old custom of dowry. One day, he decides to yield to temptation; fear of consequences had kept him away from reaching out for the 'illegal booty' of bribery which had always been within his reach. He was, as Premchand brilliantly puts it, like the traveller who sleeps comfortably all day long under the shade of a tree and wakes up in the evening to behold despairingly a mountain that he must climb before nightfall. His indiscretion lands him in jail, and his family into a cycle of misfortunes. Saman, his young and beautiful daughter who is the central character of the novel, is virtually reconciled to her fate, even gets married to a clerk drawing a most modest salary of Rs.18 per month for 'whom she had neither respect nor affection.'

Her husband's ill temper eventually drives her to despair. Not surprisingly, in such a state of mind, Saman allows herself to be seduced by the glamorous lifestyle of a neighbour, Bholi, to leave her home and find comfort, happiness and glamour in the courtesans' quarter. She would have probably stayed there but for her infatuation for a charming young client, Sadan. However, circumstances do not let her consummate her love for Sadan. In any case Saman had always been something of a misfit in the environment of a brothel. Even though it provided her with a life of comfort and pleasure.

An unexpected zeal for social reform on the part of her society elders delivers Saman from what she in any case regarded as a life of sin. Temporarily, she finds shelter in an *ashram* or home for widows but here too she is not allowed to live in peace, because her family particularly her sister had to suffer the consequences of her fall into sin which added to her sense of misery. In the end, virtually forsaken by all, Saman is installed as a mistress in an orphanage. The first part of the book deals with Saman's descent into a life of sin and the second with her eventual escape from it.

The second part of the novel is also marked by a long debate on the question of the removal of the courtesans' quarter from its place in the city. It is quite hilarious and involves many respectable elders of the community, almost each one of them with a secret agenda of his own, and brings to surface the hyprocrisy of most of them. What Premchand has depicted was actually a phenomenon of his age: in the thirties and the forties there was almost a national movement for the shifting of brothels from their traditional locations, and for long it was debated whether this would amount to getting rid of evil or create a host of social and economic problems. Incidentally, Ghulam Abbas, an eminent Urdu writer of the 1940s, wrote his famous novel *Anandi* based on the same theme and almost appears to have been inspired by the debate initiated by Premchand in *Bazaar-e-Husn*.

Premchand never allowed his debate to become too academic and polemical; with his remarkable skill as a storyteller he made it both serene and hilarious.

While steering away from sermonizing, Premchand's novel on the whole stays close to the criterion laid down by him in his presidential address at the PWA in 1936: that great literature should not aim at being merely 'beautiful' but also be rooted in a commitment to the emancipation of the society and a crusade against exploitation of man by man. Most important of all, nowhere in the story does the interest of the reader flag.

In her translation of *Bazaar-e-Husn*, Amina Azfar has done a commendable service both to the Urdu language and to Munshi Premchand. Her *Courtesans' Quarter* has the same flow, simplicity and impact as the original. The glossary of Urdu words, explanatory notes and details of Premchand's literary career provided by her does a great deal to add to the reader's interest in the great work of a great writer.

M. H. ASKARI

A Chronology of Munshi Premchand

1880   Born Dhanpat Rai, at Benaras, India. His father Munshi Ajaib Lal came from the district, Paande Pur, located near Benaras.

1895   His father, a clerk in the postal department, dies, and Premchand begins to fend for himself and his family.

1900   Writes a novel, *Prema* in Hindi.

1901   Begins to write articles for the magazine 'Zamana' which is published from Kanpur.

1904   Matriculates from the Collegiate School, Benaras.

1907   Publishes a volume of patriotic short stories entitled *Soze-Vatan* which provokes the wrath of the government. Premchand is forced to apologize and surrender 500 copies of his book which are then burnt under the Collector's orders.

1912   *Jalwa-e-Isaar.*

1917   *Bazaar-e-Husn* (*Courtesans' Quarter*).

1920   Withdraws from government service under which he had advanced to the post of Deputy Inspector of Schools. Joins the Non Co-operation Movement, under the influence of Gandhi.

1922   *Goshae Afiyat.*

1922   *Chogan-e-Hasti.*

1928   *Nirmala.*

1929   *Bewa.*

1932   *Maidaan-e-Amal.*

1936   *Gaudaan.*

1936   Presides over the first conference of Progressive Writers.

1936   Dies.

Principal Characters

DAROGHA KRISHAN CHANDR, a police officer

GANGAJALI, his wife

SAMAN, their elder daughter

SHANTA, their younger daughter

MAHANT RAM DAS

BANKA BIHARI

CHAITOO

PANDIT OMA NATH, Gangajali's brother

JANHVI, his wife

GUJADHAR PARSHAD (also Gujadhar Panday, Gujadhar Pande and Pandit Gujadhar Parshad), Saman's husband

BHOLI, a courtesan

PADAM SINGH, a lawyer

SOBHDRA, his wife

BABOO BITHAL DAS, old friend of Padam Singh

MUNSHI ABUL WAFA, Muslim member of the Municipal Board and manager of a perfume factory

ABDUL LATIF, Muslim member of the Municipal Board and a big landlord

KAPATAE, guest at Padam Singh's function

RUDRA, guest at Padam Singh's function

JITAN, Padam Singh's servant

NAZEERAN, a grocer

ZAHOORAN

JIYORI

SETH BALBAHADAR DAS, city notable and Hindu member of the Municipal Board

MADAN SINGH, Padam Singh's brother

BHAMA, his wife

SADAN, their son

MUNSHI BEJA NATH, co-owner, with Madan Singh, of the village in which they lived

SETH CHAMAN LAL, city notable and Hindu member of the Municipal Board

DR SHYAMA CHARAN, city notable and Hindu member of the Municipal Board

KANWAR ANARUDH SINGH, city notable and Hindu member of the Municipal Board

PANDIT DINA NATH, city notable and Hindu member of the Municipal Board

HAJI HASHIM ALI, leader of the Muslims in the city and member of the Municipal Board

SYED TEGH ALI, a Shia Muslim, trustee of an Imambargah (see explanatory notes), and member of the Municipal Board

RUSTUM BHAI, Parsi member of the Municipal Board

ROMESH DUTT, a college professor and member of the Municipal Board

LALA BHAGAT RAM, a contractor and Hindu member of the Municipal Board

PANDIT PARBHAKAR RAO, Editor of a Hindi language paper and a Hindu member of the Municipal Board

SYED SHAFQAT ALI, Muslim member of the Municipal Board

HAKIM SHUHRAT KHAN, Muslim member of the Municipal Board

SHARIF HASAN, Muslim member of the Municipal Board

SHAKIR BEG, Muslim member of the Municipal Board

PART ONE

There are times in one's life when one regrets one's own decency. This was such a moment in the life of Darogha* Krishan Chandr. In his twenty-five year career his character had remained unsullied by greed. Even at the age when the heart yearns for the trappings of wealth he had managed to withstand temptation. Yet now he was lamenting this very probity and proud self sufficiency. His wife, Gangajali had guarded him from temptation, but now she too was worried, and offered him no reassurance.

Darogha Krishan Chandr was a courteous, pleasant, and cultivated man. His behaviour towards his subordinates was friendly and fraternal. But his subordinates did not appreciate these qualities. They complained that he denied them the opportunity to make gains, that his gentleness and courtesy were no substitute for power and riches. They coveted the choice morsel, even if it meant putting up with insults.

His officers and bosses were not pleased with him either. When they toured other police stations they were greeted with lavish hospitality. The officers in charge, clerks, and peons partied for weeks. Offerings were made to the officers in charge. The peons were given tips, the high officials received bribes. But no such largesse was ever distributed in Krishan Chandr's domain. This austerity was construed as rebelliousness.

But despite his integrity, the Darogha was not a thrifty man. He was frugal in his personal expenditure but felt it was his duty to provide generously for his family. His was a small household consisting of his two daughters, his wife, and himself. But he spent all his earnings on them. Inevitably, he fell prey to the allure of elegant fabrics, expensive perfume, jewelled writing cases, or rugs from Agra, whenever he saw these in the market.

Gangajali was a prudent woman. She used to counsel him to be less extravagant. She would say, 'You have to find the money to marry off your two daughters. If you don't put it away now there is nobody who will help you when it is needed.' But he never took such advice seriously. He would brush it aside saying 'It will be taken care of. After all, nothing has been held up for lack of funds, so far.' But sometimes he would respond irritably, 'Don't say such things. They make me anxious.' And so the days passed.

Meanwhile the two girls, nurtured in comfort, were blossoming like lotus flowers. The older girl, Saman, was dainty, playful, lively, proud, and elegant. The younger daughter Shanta was soft spoken, innocent, and of a serious disposition. Saman always desired what was better. If two identical saris were bought for them she would turn away. Shanta on the other hand was content with whatever came her way.

Gangajali was a conservative woman. She wanted her daughters to be married early. But the Darogha would evade the issue saying, 'They are still too young. It says in the Scriptures that it is not right to marry before the age of sixteen.' Perhaps he thought that he would receive help from the Unseen. Whenever he read in the newspapers that a resolution had been passed on the abolition of dowry,* he was very pleased. He would tell Gangajali, 'This vulgar custom will end in a couple of years. There is no need to worry.' Soon Saman had entered her sixteenth year.

The Darogha could deceive himself no longer. His nonchalance did not arise from the confidence which is the result of a realistic appraisal of one's means; its roots lay in his easygoing attitude. He was like the traveller who sleeps comfortably all day long under a shady tree but wakes up in the evening to behold despairingly, a mountain that he must climb before nightfall. The Darogha was beside himself with anxiety. He began a frenzied search for a bridegroom. He sent for the horoscopes.

He desired educated families for his daughters to marry into, because he thought that among such people the question of money would not arise. But to his surprise he learned that the price of a bridegroom was in proportion to his education. (The more educated the candidate, the higher the dowry). When a horoscope was found to be favourable and it was time to settle the details, Krishan Chandr would feel his world grow dark as he heard the demands. Some asked for four thousand rupees, others wanted five thousand, and there were those whose demand was even higher. He would then come home in despair.

He remained in this state of unabated agitation for several months but found no solution to the problem. It is true that the educated parties sympathized with him, but their justification for the stand they were taking rendered him speechless. One gentleman told him, 'I detest these vulgar customs, but what can I do? Only last year I had to marry off my daughter. Her dowry was two thousand rupees, and I had to find two thousand more for the wedding feast. Tell me, how else can I make up for this loss?'

Another gentleman had this to say, 'I spent a fortune on my son's education, from which your daughter will benefit as much as he will. So don't you think you ought to share this burden of expense with me?' To such logic Krishan Chandr had no answer.

When all his efforts ended in such experiences, Krishan Chandr lost hope. He began to see his own honesty and probity as vices and he thought ruefully, 'Alas, had I been cleverer I would not have been spurned in this way.' After remaining silent for a long time he spoke out, 'See, how little use the world has for integrity and honesty. If I too had robbed society to fill my coffers, people would have come running for my daughters' hands. Since I did not, it appears that I am not even entitled to ordinary politeness. Perhaps even God hates integrity. It seems I have only two options: either I marry my daughter to a beggar, or find wealth. The first is unthinkable, so I must look for wealth. I have tasted the fruits of integrity. The experience has taught me a bitter lesson. Now I will take bribes; there is no other way. That is what the world wants, the nation wants, and maybe, that is what God wants as well. I am innocent but I am being forced into sin. I am being pressed into the service of avarice so that I can harass the poor! So be it. I will do what others do.'

With her head bowed, Gangajali was listening to her husband's anguished speech. There were tears in her eyes and her tongue was still.

– 2 –

In the Darogha's jurisdiction there lived a certain Mahant* Ram Das. He was the Mahant or head of a Hindu brotherhood of ascetics. He conducted his business in the name of Banka Bihari—the dandy Bihari. Banka Bihari made the business deals, and never settled for an interest rate of less than thirty-two rupees on every hundred that he lent. He supplied grain for sowing and made one and a half rupees on every one rupee worth of seed. He was the one who filed the lawsuits, he had the mortgage deeds documented. Nobody dared cheat Banka Bihari of his money, nor did anyone have the courage to demand his own money back from him. They knew that if they annoyed him it would become impossible for them to live in the locality.

Ten or twenty sadhus or holy men lived permanently with Mahantji. They were coarse, stout young men who worked out in the arena, drank fresh buffalo milk in the morning, strained milky hemp in the evening, and smoked pot all day. Who could dare to take on such a gang? Mahantji had his contacts among the officers too. Banka Bihari fed them consecrated sweetmeats. Their sanctity was not to be denied. Even God followed the men of the world when He came down to earth.

When Mahantji came out to survey his territory, he advanced in a procession of pomp and splendour. Banka Bihari rode ahead. Behind him was the palanquin in which sat Mahant Ram Das. In the rear was a dazzling army of ascetics displaying banners that had the name of Rama, the Hindu god, blazoned on them. Camels were loaded with tents and colourful canopies, bullock carts carried other paraphernalia. This army caused havoc in whichever village it was headed for. This year Mahantji had been on a pilgrimage. When he returned, he held a banquet. Five thousand ascetics were invited. Preparations were made for months in advance. For this occasion a contribution of five rupees for every plough was extracted from every tenant farmer. Some donated cheerfully, some had to borrow in order to comply, and some wrote out bonds; but none could disregard a command from Banka Bihari. None, that is, with the exception of a man by the name of Chaitoo. This man was old and penniless. He had had bad harvests for many years. And to top it all, only recently Banka Bihari had brought a suit against him for enhancement of revenue, burdening him with more debts.

Chaitoo refused to contribute. He did not even produce a promissory note. Banka Bihari could not tolerate such rebellious behaviour, so one day several disciples went and fetched Chaitoo, and there in front of the temple they battered him. Since he could not retaliate physically, he lashed out at his tormentors with his tongue until that tongue was silenced. That night Chaitoo died and early the next morning his death was reported to the police.

Darogha Krishan Chandr felt that a golden bird had alighted on his shoulder. He started an investigation but found that nobody was ready to give evidence in a law court even though people did come up to him privately to give him their account of the murder.

Several days passed. At first Mahantji stayed proudly aloof. He was certain that the secret would never be out. But when he heard that several persons had yielded to the Darogha's questioning he decided to start the process of negotiation in secret. He sent his agent to the Darogha, and talks leading to the transaction began. Daroghaji said, 'My temperament is well-known. I consider a bribe to be no better than a poisonous snake.' The agent replied 'Yes we know that. But you must make some allowance for holy men.' After this the two men talked in whispers. The agent was heard to say, 'No sir, five thousand rupees is too much. You know Mahantji. He is so stubborn that he'd rather go to the gallows than give an inch. If you settle on a reasonable sum, he will not be distressed, and you will stand to gain.' Finally, the figure of three thousand was agreed upon. But buying a bitter pill, and bringing oneself to swallow it are two different things. The agent went back to Mahantji to inform him of the outcome of the negotiations, but Krishan Chandr began to think 'What am I doing?' On the one hand was wealth, and freedom from anxiety, on the other the destruction of one's conscience and fear of consequences. He had neither the pluck to give in to his temptation nor the strength to turn it down. After a lifetime of delicacy and cautiousness, he was miserable at the thought of betraying his principles. He thought, 'If this is what I was going to do in the end, why didn't I begin twenty-five years ago? I would have amassed a fortune in gold and property by now. At the end of a contented life spent in frugality, why did I have to inflict this stigma upon myself?'

But then his wounded self rose to its own defence. 'It's no fault of yours. You didn't lower your principles merely to buy yourself more luxuries. But if the culture of your land, the greed of others, and a sacred duty* have forced you to leave the straight path, you are not to blame. Your conscience is intact and you are still innocent in the eyes of God.' And with such logic Daroghaji consoled himself.

But the second stage was yet to come. Fear of the consequences had always kept Krishan Chandr from reaching out for illegal booty. So he had never acquired the nerve to make such a move. A man who has never raised his hand at another cannot suddenly become capable of striking him with a sword. And so the Darogha was plagued with the thought of what might happen to him if his secret were divulged. Imprisonment would become an inevitable. And his good reputation of a lifetime would be mud. But while it is possible to appease the conscience with logic, the dread of consequences cannot be set at rest in the same way. Such fears need secrecy; and so the Darogha tried his best to keep the matter secret. The agent was told that on no account should anybody get wind of the business. It was concealed also from the constables and the staff.

It was nine in the evening. Daroghaji had sent his three constables out on some pretext. The watchmen had also been dispatched, to buy provisions. Darogha was sitting alone, waiting for the agent who had still not returned. 'Why has he not come?' the Darogha thought. 'What if the watchmen return? I had told him to come soon. What if Mahant does not agree to give three thousand? I'll not take a penny less.' Daroghaji began to calculate how much of the money he would pay as the dower and how much he would spend on the wedding feast. After about half an hour the agent arrived. Darogha's heart was beating hard with hope and apprehension. He got up from the string bed and, affecting nonchalance, began to stain a betel leaf.* Meanwhile the agent came in.

Krishan Chandr: Well?

Agent: Mahantji …

Krishan Chandr looked in the direction of the door and said 'Have you brought the money or not?'

Agent: Yes, I have brought it, but Mahantji has….

Krishan Chandr again looked all around, then said, 'I will not take a penny less.'

Agent: Yes, but you will give me my cut, won't you?

Krishan Chandr: You can take it from Mahantji.

Agent: I charge five per cent.

Krishan Chandr: You won't get a penny from out of this. I am murdering my own conscience, not usurping any body's rights.

Agent: As you like, but you are depriving me of my right.

The ox drawn carriage was made ready and Krishan Chandr mounted it. He would have liked to fly home. Again and again he told the coachman to hurry. Finally at about eleven o'clock he reached home. Gangajali had been waiting. She said, 'Why are you so late?'

Krishan Chandr: Something came up, and the place is far.

After dinner Daroghaji lay down but he could not sleep. He was too embarrassed to tell Gangajali about the money. She in her turn stole glances at him as though trying to fathom his innermost thoughts. Finally Krishan Chandr asked her, 'If you were standing by a river and a tiger rushed at you, what would you do?' Gangajali understood his meaning and replied, 'I would jump into the river.'

Krishan Chandr: 'If your house were on fire and the doors were locked, what would you do?'

Gangajali: I would jump from the roof.

Krishan Chandr: Did you understand the meaning of my questions?

Gangajali: I am not a fool.

Krishan Chandr: I have jumped. I don't know whether I will sink or swim.

– 3 –

Krishan Chandr did not know how difficult it is to swallow forbidden fruit. He was a novice in the art of bribe-taking. He had observed secrecy not in order to deprive anybody of a share in the booty, but because he feared for his reputation.

When he left Daroghaji, the agent went straight to the police station and there he let the cat out of the bag. The staff exclaimed, 'So, he is trying to pull the wool over our eyes! As though we did not work for the Government. We'll see how he hangs on to his prize. Wait till we expose him!'

Ignorant of what was brewing, Krishan Chandr was busy trying to arrange his daughter's marriage. Negotiations were in progress with an affluent family where the marriage had been proposed. There were comings and goings between the two families, as well as parleys and discussions. Meanwhile, not far away secret letters were being dispatched to officials. In these letters the matter had been laid bare. The officials made secret investigations of their own and the facts of the case became crystal clear.

A month had passed and the first ceremonies in connection with the wedding were about to begin. Daroghaji was in the police station, stretched out on a divan. When he saw the police superintendent coming in with several constables and two police station officials, Krishan Chandr sat up with a jolt. He had no inkling that his officers had made secret investigations and had unearthed all the facts. They had even completed the preparation of evidence for the prosecution.

One sub-inspector produced a warrant for arrest and showed it to Krishan Chandr. The Darogha turned pale. He was stunned, and stood with his head bent. His expression showed no fear, only regret. These were the very same two sub-inspectors before whom he used to walk with his head held high; whom he regarded with disdain. In a moment the good reputation of a lifetime had been shattered. His inner self told him, 'Suffer for your sins. Didn't I warn you not to jump into this inferno? But you didn't heed me. Had you been content to marry your daughter into a family of modest means you would not have seen this day. You could think of nothing but glory and status, so now eat the fruit of this evil transaction.'

The superintendent asked him, 'Krishan Chandr, do you have anything to say?'

Krishan Chandr thought, 'Shall I say that I am blameless, that this is a conspiracy hatched by my enemies who, tired of my integrity, have resorted to this ploy?'

But he could not bring himself to be so brazen. He was still a novice in this field. His feelings of guilt had degraded him in his own eyes.

Just as the unscrupulous rarely have to pay for their sins, so the well intentioned seldom escape punishment. Their faces, their eyes, their movements, all seem to turn into tongues that testify against them. Their morality becomes its own judge. He who is used to travelling on the straight path is bound to lose his way if he chooses a devious course.

The superintendent again asked, 'Is there something you'd like to say for yourself?'

Krishan Chandr replied, 'Yes. I want to say that I have sinned and that I should not be spared. Blacken my face and have me paraded through the town. Push me into a blazing pit of fire. Because, in order to achieve spurious honour and success I committed this immoral act, I want to be punished for it. The whip of integrity could not keep me on the straight path so I am only fit for the chains of justice. Let me go home for a while only, and then I will be ready to accompany you.

While he was talking, Krishan Chandr felt not only repentance but also the relish of pride. He wanted to show the two officers that if he had committed a crime he could also endure the consequences bravely. He would not shirk them.

They in turn listened to him in astonishment. They stared at each other as if to say, 'Has this man gone mad? If he wanted to be so moral why did he commit this crime in the first place?'

The superintendent who was looking compassionately at Krishan Chandr, gave him permission to go home.

Gangajali was sitting with a silver platter in front of her. On it she was placing the colourful accoutrements for the tilak ceremony.* Krishan Chandr told her, 'Ganga, the secret is out. I have been arrested.'

Gangajali looked at him in astonishment. She grew pale and tears fell from her eyes. Her worst apprehensions had come true.

Krishan Chandr said, 'Why do you weep? They are not treating me with injustice. I am being punished for my deeds. They will probably bring a criminal charge against me. But don't worry, I am ready for any punishment. This retribution will cleanse the ill-gotten money, and you should use it for the girls' marriages. Don't spend a penny of it on lawyers and agents for me. It will grieve me if you do. Having destroyed my conscience, my reputation, and my life, I should at least have the satisfaction of knowing that I have done my duty by my girls.'

Gangajali beat her head with both her hands. She was so furious with her shortsightedness that she thought, 'May my body be set on fire and I burn to ashes.' She felt on her heart, the heat of strong sunlight arising from behind a cloud of grief and regret. She looked at the heavens with despair in her eyes. 'If only I had known that it would come to this, I would have married my daughter to a penniless man, or poisoned her!' She leaped up and grabbing Krishan Chandr's hands spoke wildly, 'Burn that money. Throw it on the head of that bloodthirsty Ram Das. My girls will remain unwed. Oh God, why was I such a fool! I will go to your boss myself. I am not ashamed. This is no time to hold back.'

Krishan Chandr: Nothing can be done now.

Gangajali: No! take me to the Sahib. I'll fall at his feet and tell him, 'Here is your money, take it, and if you must punish somebody, punish me. I am the root of the mischief. I am the one who sowed this seed of sin.'

Krishan Chandr: Keep your voice down. They can hear you outside.

Gangajali: Why don't you take me to the Sahib? He will have mercy on a helpless woman.

Krishan Chandr: This is no time to shed tears. I am caught in the clutches of the law and cannot escape. Be patient. If God wills it we will meet again. Saying this Krishan Chandr prepared to leave. The two girls clasped his legs and Gangajali grabbed his waist with her two hands. There was a commotion and Krishan Chandr too was

overwhelmed. He was thinking 'What will become of them. Oh God,
You are the protector of the helpless. Look after them.'

In a moment he had pulled himself free from his wailing family
and left the house. Gangajali had thrust out her hands to grab him
again, but her hands remained spread like the wings of a wounded
bird.

– 4 –

The men in Krishan Chandr's department mistrusted him, but he
was universally loved among the people in his town. The news of
his arrest caused an uproar among his townsmen. Many notables came
forward to offer bail, but the superintendent rejected their offers.

A week later the cases were presented before the subdivision
administrator. The hearing lasted for two months after which they
were transferred to the sessions court where the proceedings continued
for a month. Krishan Chandr was sentenced to five years of
imprisonment with hard labour, and Mahant Ram was given seven
years' imprisonment. Two of the Mahant's disciples were given life
terms.

Gangajali had a brother named Pandit Oma Nath. He had never got
along with Krishan Chandr who used to call him a deceitful rival and
ridiculed his long tilak.* For this reason Oma Nath rarely visited him.
But shaken by the news of his arrest he came and took his sister and
two nieces home with him. Krishan Chandr himself had no brother,
only two cousins, sons of his paternal uncle; but they showed no
concern.

When he was leaving, Daroghaji had forbidden his wife to spend
even a single penny from Ram Das's money on lawsuits for his
acquittal. He had been certain that he would be let out. But Gangajali
could not be patient, and ignoring his admonition, she spent the money
recklessly. Until the last moment the lawyers kept assuring her that
Krishan Chandr would be set free.

The judgement was appealed in the Supreme Court, where
Mahantji's sentence was maintained, but Krishan Chandr's was
reduced by one year.

Gangajali had returned to her paternal home but soon regretted her
move as a bad one. This was not the home where she had spent her
childhood, where she had played with her dolls and built little mud
houses for them. Her parents had died and she saw no familiar faces in

the village. Even the fields had been sown with trees and the old trees had been felled and were replaced by fields. She recognized her old home with difficulty, but the worst affliction that she had to bear was that there was neither affection nor respect for her in her brother's home. Her sister-in-law, Janhvi, was always hostile to her. Janhvi had no interest in her own home; she used to spend her time at her nieghbours', where she moaned and groaned over her misfortune in having to put up with Gangajali. She had two daughters who also kept their distance from Saman and Shanta.

Gangajali had spent all of Ram Das's money. Only the four or five hundred rupees which she had carefully saved from her household expenditure, remained. Therefore it embarrassed her to approach the subject of Saman's marriage with Oma Nath. Meanwhile six months had passed, and the engagement had been broken by the boy's people.

However Oma Nath was deeply concerned. Whenever he had some time, he would go away for a few days to look for a suitable husband for Saman. As soon as he arrived in a village, there would be a great stir. People would bring out from their bundles, clothes that they ordinarily wore at weddings, and put them on. They would wear borrowed jewellery. Mothers would bathe their children, put kohl in their eyes, and send them out to play wearing clean clothes. Old men who wanted to get married, went to barbers to have their moustaches trimmed and their white hair plucked. As long as Oma Nath stayed in the village women did not leave their houses, they did not go out to fill water, nor did they work in the fields. But Oma Nath was not taken in by this deception. Saman was so beautiful, so accomplished in the household arts, and so well educated. Her life would be ruined in the home of these rustics, he thought.

Finally Oma Nath decided to look for a bridegroom in the city. But when he heard the demands of the city people he was stupefied. Even the clerks and scribes asked for figures in the thousands. People shied away when they saw him. A few persons showed willingness on the basis of his fine pedigree, but either their horoscopes did not match or Oma Nath himself was not satisfied.

A full year passed in this way. Oma Nath was tired of running around. He was like a distributor of advertisements, who, having distributed advertisements to well-dressed people only, all day long, and finding in the evening that he still has a pile of material left over, decides to forego the condition of sartorial elegance, and dispenses his goods to all and sundry in order to get rid of his burden. Oma Nath now retained only one condition—that of good breeding. For him good breeding was the most desirable attribute of all.

It was the month of Magh.* Oma Nath had gone to bathe in the Ganges.* When he finished, he went straight to Gangajali and said, 'Sister, the marriage has been arranged.'

Gangajali: Your hard work has borne fruit. The boy, is he a student?

Oma Nath: Not a student, he is a clerk in a factory where he earns fifteen rupees per month.

Gangajali: What's his age?

Oma Nath: About thirty, I should think.

Gangajali: Is he good-looking though?

Oma Nath: Nobody is bad-looking in the cities. Everybody has nice hair and white clothes. His name is Gujadhar Parshad.

Gangajali said in a disappointed tone, 'If you like the match, I like it too'.

Oma Nath had already made the arrangements, so the wedding took place in the month of Phagun.* When Gangajali saw her son-in-law she felt as though her breast had been stabbed with a spear, or as if her daughter had been pushed into a well.

When Saman came to her husband's house she found that matters were even worse than she had thought. The house consisted of two small cabins and one shed. Salinity had seeped into all the walls and the plaster was peeling off. The stink from the open drains outside came in. There was no sunlight. And for this house a rent of three rupees had to be paid every month.

However, Saman spent the first two months quite comfortably as Gujadhar's old aunt used to do all the housework. But in the summer cholera spread in the city, and the old woman succumbed to it. Gujadhar was very worried. The part time help demanded three rupees a month for doing the washing up. For two days the stove was not lighted. Gujadhar could say nothing to his wife. Both days he bought puris* from the market. He wanted to keep Saman happy because he was smitten by her beauty. On the third day he woke up an hour before sunrise and finished all the washing up, cleaned the dining area, and filled the water pots. When Saman woke up she was amazed to see the transformation. She understood at once who had done the work but was too embarrassed to ask her husband. In the evening she did everything herself, weeping as she washed the dishes. But in a few days she became accustomed to doing the housework, so much so that she began to feel a certain satisfaction in her life.

Gujadhar was ecstatic. He went about praising Saman to his friends. 'She is not a woman, she is a goddess. She comes from a great house but does even the tiniest chores with her own hands. She cooks so well that I am never tired of eating,' he told them. The next month, when he was paid his salary he put it all in Saman's hand. That day Saman experienced the pleasant sensation of freedom. 'Now I will not have to plead for every penny. I can spend this money as I like. I can eat what I like' she thought. But not being adept at the art of housekeeping, she could not tell the difference between the important and the unimportant uses of money. The result was that ten days before the month was over she had spent all the cash that she was given. After all, she had been taught calligraphy, but not the management of a home.

When Gujadhar heard the news, he was stunned. The question of how to meet expenses the rest of the month overwhelmed him. He had had some inkling of this earlier. However he said nothing to Saman, though he suffered from intense anxiety all day. 'Where will the money come from?'

Although Gujadhar had made Saman the mistress of the house, he was by nature very grudging. Jalaibis* at breakfast tasted like poison to him. When he saw ghee in the lentils he felt a constriction in his chest. At dinner he watched the batlees* closely, afraid that too many had been cooked. If he saw lentils or rice scattered wastefully by the door he fumed inwardly. But he was silenced by Saman's maddening beauty.

However, on that day, after he had unsuccessfully touched several people for a loan he became impatient. When he came home he told Saman, 'You've spent all the money. Now tell me where I can get more?'

Saman: I haven't squandered it!

Gujadhar: But you knew that this was all we had.

Saman: Such a small amount can't be stretched.

Gujadhar: Well, I cannot resort to robbery!

But Saman had been brought up in ease and comfort. She was used to eating well and dressing well. When she heard the hawkers' call she became restive. So far she had been including Gujadhar in her feasts, but now she learned to gorge herself, alone. To satisfy her palate she began to deceive her husband.

– 5 –

Gradually the neighbourhood came to hear of Saman's beauty. The women of the locality began to visit her. Saman used to despise them, so she was not open with them. She had the style that distinguishes nobility. Very soon her neighbours came to accept her dominance. Saman appeared like a queen among them and her vain temperament derived pleasure from this. She used to display her talents before them. The poor women would complain of their fate before her, while she extolled her own. When they indulged in backbiting, she stopped them. She used to sit before them in her silk saree which she had brought in her trousseau. She would hang her silk jacket on a peg. This spectacle made an even greater impression on her visitors than Saman's genteel manners. They soon began to take her views as the last word in the matter of jewellery and clothes. When they ordered new jewellery they sought her opinion. When they bought new sarees they made it a point to show their purchases to Saman. Outwardly Saman's manner was objective when she was advising them, but inwardly she was grieved. She would think, 'All of them buy new jewellery and new clothes while I don't know where my next meal is coming from. Am I the most unfortunate being in this world?'

Gujadhar was working extremely hard those days. As soon as he returned from the factory he would go to a Hindu banker's to do his accounts. From there he came home at eight in the night. For this extra work he was paid five rupees. But in spite of the increase in his income he could see no change in his financial status. All his earnings were spent on food. His cautious nature was perturbed by such extravagance. But to crown it all, Saman would curse her fate before him, causing him to become even more agitated.

He could see clearly that Saman cared for him less and less every day. What he did not know was that Saman pined more for the delights of the palate than she did for the pleasures of the heart; that she found sweetmeats sweeter than love's murmurs; that she found colourful clothes more delightful than a lover's feigned tone of grievance.

No matter how secure we are in our habits, they are bound to be influenced by the company we keep. Saman may have had much to teach her neighbours, but she learned a great deal more from them herself. How casual we are about our domestic life! We think that we need no preparation or education for it. A child who plays with dolls, or a young girl who plays with her friends is considered fit to be the mistress of a household. On a playful calf is loaded a heavy burden of grain! The women with whom Saman spent her time used to consider their husbands merely a tool in the service of their person. According to them it was the duty of the husband to keep them decked up with jewellery and well-fed on sumptuous food. If he could not do all this he was a nincompoop, a wastrel. He had no right to marry. He did not deserve love or respect. This was what Saman learned from them. So, whenever Gujadhar was angry with her, he would have to listen to her long speeches on the duties of a husband.

There was no dearth of debauchees and licentious young men in the neighbourhood. On their way to school, boys would stare at Saman's door. When the rakes passed by her house they would sing love songs. No matter how busy she was, when she heard their voice Saman would come and stand behind the cane blinds. Her bold nature enjoyed such coquetry. She would do this merely to show off her beauty, and provoke others.

– 6 –

Facing Saman's house was that of a prostitute named Bholi. Bholi would sit at her window on the first floor, adorned with ornaments and cosmetics. The sound of melodious singing came from her room until late into the night. Sometimes she went out for a drive in her phaeton. Saman used to despise the prostitute.

Saman had always heard that prostitutes were contemptible and immoral; that they entrapped young men with their coquetry. No decent man ever spoke to them. Only the licentious went to see them secretly at night. Many times Bholi motioned her to come over when she saw her standing behind the cane blinds. But Saman thought it beneath her dignity even to speak to her. She said to herself, 'I may be poor but at least I am virtuous. I am not barred from the homes of decent people. Bholi may be living in luxury but she is not respected. All she can do is to sit in her brothel and expose her shamelessness.' But very soon Saman realized that she was wrong in despising the prostitute.

It was the month of Asarh. Finding the heat intolerable, Saman lifted the cane blinds towards evening, and sat fanning herself by the open door. In front of Bholi's house preparations were being made for some function. Water carriers were sprinkling water. A marquee was being hoisted in the yard. Carts loaded with lights and chandeliers were arriving. Seating arrangements were being made on the ground.

Dozens of people were running about, busy with the arrangements. On happening to see Saman, Bholi came over and said, 'There is a maulood at my place. If you want to come I'll have the curtain put up.'

'I'll see it from where I am.' Saman replied nonchalantly.

Bholi: You will see it but you won't be able to hear it. Let me have the curtain put up, there is no harm.'

Saman: I am not so keen to hear it.

Bholi looked at her compassionately and thought, 'This peasant has probably just come from the village. God knows what she thinks. Well she'll find out today who I am.' After that Bholi did not press her invitation.

It was night. Saman looked at the stove with a shudder. Willy nilly she got up, lit the stove, put some rice and lentils on it, to cook khichri, and came back to her post to watch the spectacle. By eight o'clock the gas lights had turned the canopy into a brilliant dome. Flowers and greenery enhanced the elegance of the venue. Spectators began to appear from all directions; some came on bicycles, some on horse drawn carriages, and some on foot. After a short while two or three phaetons also arrived. In about an hour the whole yard was full of people.

Finally the maulana came; he had a bearing that inspired awe. He sat down on the decorated couch, and the maulood began. Several people were looking after the guests. Some sprinkled rosewater over them, others passed the khasdan, containing betel leaf. Saman had never seen such a gathering of the elite.

At nine o'clock, Gujadhar Parshad returned home. Saman gave him his food. After dinner Gujadhar also left for the function. But Saman had no appetite. She sat watching the proceedings until eleven o'clock. After that, portions of sweetmeat were distributed among the guests and at midnight the function ended. When Gujadhar came home, Saman asked him, 'Who were those people?'

Gujadhar: I don't know all of them. There were all sorts, including some of the city's elite.

Saman: Don't those people find it demeaning to go to the house of a prostitute?

Gujadhar: Why would they go if they thought it demeaning?

Saman: But you must have felt ashamed to go there?

Gujadhar: Why should I feel ashamed when so many of the elite were present there. The Seth for whom I work in the evenings was also there.

Saman said thoughtfully, 'I used to think that people found such women contemptible.'

Gujadhar: Yes, there are such people, but now they are very few. Western education has emancipated them. Bholi is much respected here.

The sky was overcast and there was no breeze. Gujadhar, exhausted after a long day, was fast asleep as soon as he lay down on the string bed. But Saman lay awake.

The next day when she again lifted the blind and sat down, she saw Bholi sitting on her terrace.

She came out on the verandah and addressed Bholi herself. 'You had a lively function last night' she said. Bholi understood that she had scored a victory. She smiled and said, 'Shall I send you some of the sweetmeats? They were made by the confectioner and a Brahmin brought them over.' Saman replied shyly, 'Yes, please send some.'

– 7 –

It was a year and a half since Saman had come to her husband's house but she had not been able to visit her family. They wrote her letters, and when she replied she would reassure her mother saying, 'Don't worry on my account, I am very comfortable.' But now her letters were full of her troubles. She would write, 'I spend my days in weeping. Why did you push me into this dark well?'

Saman had stopped praising her family to her neighbours. There was a time when she used to boast about her husband to them, now she did nothing but complain about him. She would tell them, 'Nobody cares for me. My family thinks I am dead. My mother's house has every comfort but what good is that to me? She thinks that I am living in the lap of luxury here. Little does she know how I suffer.'

Saman had become drier than ever in her treatment of Gujadhar. She thought he was responsible for her troubles. She would wake up late in the morning, and not sweep the house for days on end. Sometimes Gujadhar had to go to work without breakfast. He could not understand what the matter was, why things had changed so drastically.

One night when Gujadhar returned home at eight o'clock, he found the door locked and the lamps unlit. He thought, 'It has come to this. Where could she have gone at this hour?' He knocked loudly at the door so that if she were sitting at a neighbour's she would hear him and come home. He thought, 'Tonight I'll set her right!' Saman was at that moment sitting upstairs at Bholi's, chatting with her. Bholi had asked her over with much insistence, and Saman had not been able to refuse. When she heard the loud knocking at her door she stood up in confusion and ran home. She had lost track of time while talking to Bholi.

Saman opened the door, lighted the lamp and proceeded to light the stove.

She was regretting her mistake. Suddenly Gujadhar said in a wrathful voice, 'What were you doing there so late at night? Have you no shame at all?'

Saman replied in a low tone, 'She had asked me to come, many times, so I went over. Take off your clothes. Dinner will be ready in no time. You've come earlier than usual today.'

Gujadhar: You can cook dinner later. I am not hungry. First tell me why you went there without my permission? Have you decided that I am a useless fool?'

Saman: It is not easy to stay all alone in this black hole.

Gujadhar: So, are you going to socialize with prostitutes? Don't you care at all for your reputation?

Saman: What's wrong in going to Bholi's house? Many great people go there; I'm only small fry.

Gujadhar: Great people can go there. But it is scandalous for you to go there. I don't want my wife hobnobbing with whores. Do you know who the great people that go to her house are? Money doesn't make anybody great. Religion has a much higher status than wealth. You have been deceived by the crowds you saw on the day of the maulood. You should know that there was not a single decent man among them. My Sethji may be very rich but I wouldn't let him set foot in my doorway. These people are too proud of their wealth to care about religion. Their going to her house does not sanctify Bholi. I strictly forbid you to set foot in her house. If you do, it will be the worse for you.

This made a deep impression on Saman. She thought, 'How was I to know who those people were? It is true that the wealthy become enslaved by such women. Bholi herself was saying so. I was deceived.'

The idea reassured Saman. She became convinced that the people who attended Bholi's function were lustful and pleasure-loving. This made her feel better as it gave her a reason for feeling superior to Bholi.

Saman's religious faith came alive. To impress Bholi with her devotion she began to go to the Ganges every day for sacred immersion. She sent for a copy of the Ramayan from which she sometimes read out portions to her neighbours, and sometimes she recited the holy book melodiously to herself. However it was not her spirit that benefited from this exercise, but rather, her pride.

It was the month of Chait.* On Ram Nomi,* Saman went with her friends to a big temple. The temple was decorated, and brilliant with electric lights. The crowd was so large that there was not even standing room in the yard.

The notes of a song, sung in a captivating voice, floated in. Saman looked out of the window into the yard, to find that the singer was no other than her neighbour, Bholi. She was sitting at the centre of the crowd, singing to a spellbound multitude.

There were numerous distinguished people in the gathering, some with tilak on their forehead, others with ashes smeared on their bodies, and yet others in garments reddened with geru.* Many of these were people who Saman saw every day, bathing in the sacred river. She was used to thinking of them as gods of knowledge and wisdom. But they were present here, mesmerized by Bholi's singing. Bholi had but to glance at someone for that person to be transported with joy, as though he had had a mystical experience. This phenomenon affected Saman like a dart of lightning. With it, her pride was shattered. Her main support, which had helped her hold her head high, was gone; and she realized that it was not just the wealthy that bowed to Bholi, but even pious men and holy ascetics were smitten by her coquettish ways. The woman whom she had wanted to humiliate with her simulated piety was sitting here in Thakurjee's sacred temple in a position of honour, while she herself could not even command standing room.

When Saman returned home she wrapped up the Ramayan and put it away. And she stopped going to the Ganges for sacred immersion. Once again life swayed perilously for her, like a boat with a broken anchor.

– 8 –

Gujadhar Parshad was like a man sitting among thieves, holding a bag of gold. Saman's beauty over which he used to hover like a moth now seemed to him like a blazing flame. He stayed away from it. He was afraid that it might burn him up. A woman's beauty is her virtue. Without her, virtue her beauty is truly a flame, terrible and lethal.

Gujadhar had done his best to provide every comfort for Saman. He could do no more. It became his greatest concern to find another house. In his present house there was no open space, so whenever he told Saman not to sit by the cane blind, she would retort, 'Shall I perish in this cage then?' The other reason was that he wanted Saman to stop associating with her neighbours, having become convinced that it was their influence that had brought about the change in her. So he used to go looking for a house, but when he found out their rent he would come home, disheartened.

One day he returned from Sethji's house at eight o'clock, to find Bholi sitting on his bed, talking pleasantly to Saman. His lips twitched with fury. As soon as Bholi saw him she came out of the room and said, 'If I had known you were working for Sethji you would have been given a promotion long ago. Saman has only just told me you work there. Sethji is very partial to me.'

This remark served only to aggravate his suffering. He thought, 'This woman thinks I am so abject that I will approve of her intervention in promoting my interests!' He made no reply.

As soon as Saman saw the expression on his face, she was ready for the storm of indignation that she knew would break. Gujadhar did not hide his wrath. As he sat down he said, 'You are associating with Bholi Bai again. Didn't I forbid you to do so?'

Saman replied boldly, 'She is not a pariah. In status and prestige she is no less than anybody else. So why is it degrading for me to speak to her?'

Gujadhar: You are talking nonsense again. Honour and dignity don't come with wealth.

Saman: But they do come with piety.

Gujadhar: So you think Bholi is a pious woman?

Saman: That only God can tell. But pious people certainly respect her. On Ram Nomi I saw her singing in a gathering of great pundits and holy men. Nobody seemed to hate her. They were in fact very attentive. Not only did they show her cordiality but they were rapturous whenever they could speak to her. They may or may not have hated her in their hearts but overtly at least one saw Bholi overshadowing everybody else.

Gujadhar: You think those people are sincere because they wear those long tilaks? Nowadays piety is the stronghold of hypocrisy. In its sacred river lie terrible beasts who swallow innocent devotees. People are taken in by men with long matted hair, long tilaks, and long beards. But they are all swindlers, who amass fortunes in the name of religion and give religion a bad name. So of course they offer Bholi a place of honour. Where else would she be so welcome?

Saman asked innocently: Are you telling the truth or are you beguiling me?

Gujadhar gave her a look of love and replied, 'No Saman, I am not beguiling you. It is true that there are very few honest people in our country. Though the country is not yet completely devoid of them. Such people are compassionate and truthful. They always wish others well. If Bholi came to them in the guise of an enchanting fairy, they would not raise their eyes to look at her.'

Saman was silent. She was thinking about what Gujadhar had said.

– 9 –

From the following day Saman stopped standing by the cane blinds. The cries of the vendors went unheeded. The rakes passed by, singing their love songs. But no longer did they catch glimpses of a face behind the blinds. Bholi called her many times, but Saman excused herself saying that she did not feel well. Bholi then came to visit her two or three times, but Saman did not greet her with warmth.

It was now two years since Saman had come to her husband's house. Her silk sarees were old and worn. The expensive jackets were now shabby. She was no longer the queen of her dominions. Her prestige was going down by the day. Deprived of good clothes she had lost her high status. She would lie all day in her tiny room, sometimes reading and sometimes sleeping.

Lying in a closed room affected her health. Headaches became frequent, and sometimes she had fever. She suffered from palpitation and problems of digestion. Even a little work made her restless. She became weak and her face that used to be like a fresh blossom, wilted.

Gujadhar was worried. Sometimes he would vent his irritation on Saman, telling her, 'You are always so inactive. If I can't even depend on you to give me my meals on time, what good are you?' But immediately after he would regret his selfishness.

Gradually it dawned on him that Saman's ill health was due to the unwholesome air that she breathed. While only a short time ago he had seethed with rage to see her standing by the cane blinds and forbidden her from bathing in the Ganges, he would now himself lift the blinds, and urge her to go to the river. On his insistence, Saman went for her bath on several consecutive days and this seemed to do her good. She then took to bathing regularly every day. And water revived the wilting plant.

It was the month of Magh. One day, several of Saman's neighbours accompanied her for a bath in the river. On the way lay Bini Garden which housed different kinds of animals. There was also a spacious dome made of thin iron wire in which birds were kept. On the way back from the river Saman's companions decided to take a stroll in the garden. Saman herself used to go home directly after her bath, but on the insistence of her friends, she too made the digression.

For a long time she walked about, looking at the assortment of strange animals. Finally she was exhausted and sat down on a bench. Suddenly, she heard a voice call out, 'Who is this woman sitting on the bench? Get up woman! Do you think this bench was placed here for you?'

Intimidated, Saman looked back, to find the watchman standing behind her. She got up from the bench, and in order to hide her humiliation, pretended to watch the birds. She was still reproaching herself for sitting on the bench when she heard a hired carriage draw up in front of the zoo. The watchman leapt up to open the door of the carriage and two women climbed down. One of them was Saman's neighbour, Bholi. Saman hid behind a tree, and the women began to stroll in the garden. They fed the monkeys and the birds, they stood on the turtle's back, and then they went away to watch the fishes in the pond. The watchman was walking behind them like a servant. While they were looking at the fish he quickly made two bouquets of flowers and offered them to the two women. After some time the two came and sat on the very bench from which Saman had been ordered to get up, while the watchman stood behind them respectfully. Saman watched this scene with eyes that rained fire. Her forehead was wet with perspiration, her body quivered like a straw, and her heart was filled with a blaze of fury. She hid her face in a corner of her garment and wept. As soon as the two prostitutes left the garden Saman sprang up like a tigress and stood in front of the watchman. With a voice quivering with rage she said, 'You made me get up from this bench as though it belonged to you. How come you said nothing to the two whores?'

The watchman replied contemptuously, 'There is no difference between you and them.'

The effect of this sentence on Saman was like that of oil on fire. Chewing her lips she said, 'Shut up you scoundrel! A tip can make you pick up a whore's shoes. I will sit on this bench now. Let me see you dare to drag me away.' At first the watchman was intimidated, but as soon as Saman sat on the bench he leapt at her with the intention of dragging her away by her hand. But Saman got up herself, staring furiously at him, looking like the picture of wrath. Her friends who, after walking round the zoo, had gathered close to where she was, stopped to watch the scene. None had the courage to intervene.

Suddenly another carriage drove up and stopped. The watchman was still grappling with Saman when a well-dressed man climbed out of the carriage and giving the watchman a hard shove, shouted, 'How dare you touch the lady!'

The watchman turned pale with fear and immediately fell back. 'Sir, is she from your household?' he timidly inquired.

The good man replied angrily, 'Whether or not she is from my household, how dare you manhandle her. If I report you, you'll lose your job.'

The watchman began to grovel. Meanwhile, the lady in the carriage motioned to Saman to come over. 'What was he saying to you?', she asked her.

Saman: He wanted me to get up when I was sitting on the bench. But he allowed two prostitutes to sit here. Does he think I am inferior to those women?

The lady said to her, 'Such people are ready to serve anybody who gives them a few pennies. It's best not to get involved with them.'

The two women got talking. The lady's name was Sobhdra. She lived in Saman's neighbourhood. Her husband was a lawyer. The couple had been on their way home after bathing in the river, when the husband noticed a watchman wrangling with a decent woman. He had immediately decided to intervene.

Sobhdra was so captivated by Saman's looks and speech that she invited her to sit with her. The lawyer sat with the coachman and the carriage drove off. Saman felt as though she were sitting in a flying coach and was on her way to heaven.

Sobhdra was neither beautiful nor stylish, yet her manner was so amiable and open that Saman was delighted to make her acquaintance. On the way, when she saw her friends walking home, she looked at them with pride, as if to say, 'Could you ever be honoured in this way?' But besides feeling proud she also suffered some pangs of fear at the thought that Sobhdra might see her house and think her despicable for living in such a dismal place.

'How lucky this woman is. Her husband is a god! If he hadn't come, that brutal watchman would not have spared me. How courteous of him to let me sit in the carriage while he himself is perched next to the coachman.' So Saman's thoughts were running when she realized that they had arrived at her house. She said to Sobhdra shyly, 'Please have the coach stopped. That is my house.' Sobhdra had the carriage stopped. Saman glanced at Bholi's house. Bholi Bai was strolling on her roof. Their eyes met. Bholi's seemed to say, 'Oh, so we are having a splendid time!' And Saman's responded, 'Have a good look. These are people whose company you can never have, even if you die for it!'

Saman climbed down from the carriage, and turned to Sobhdra with tears in her eyes. 'Don't forget me, now that you have kindled my love. I will be looking forward to seeing you', she said.

Sobhdra: No, no. We have much to talk about. I'll send for you tomorrow.

The carriage drove off. When Saman re-entered her house she felt as though she had woken up from a pleasant dream.

Gujadhar wanted to know, 'Whose carriage was that?'

Saman: Some lawyer who lives close by. I met his wife in Bini Garden. She insisted that I sit in the carriage.

Gujadhar: So were you sitting with the lawyer?

Saman: Don't be silly. The poor man was sitting with the coachman.

Gujhadar: No wonder you are so late.

Saman: Both husband and wife are decency itself.

Gujadhar: All right, you have praised them enough. Now light the stove.

Saman: I suppose you know the lawyer?

Gujadhar: In this neighbourhood there is one lawyer. His name is Padam Singh. It must be he.

Saman: He is fair and tall. Wears spectacles?

Gujadhar: Yes, yes. It must be the same person. His house is to the west?

Saman: Is he a famous lawyer?

Gujadhar: I don't keep his accounts. I see him now and then. He is a good man.

Saman became aware that Gujadhar did not like to hear the lawyer mentioned. She changed her clothes and proceeded to cook the meal.

– 10 –

The following day Saman did not go for her bath. Instead, she sat down to mend the tear in one of her silk sarees, early in the morning. In the afternoon one of Sobhdra's maidservants came to fetch her. Saman had thought that the carriage would be sent for her. She was disappointed to see the servant; it was what she had been afraid of. She accompanied the maid to Sobhdra's house where she spent two or three hours. At the end of it she was still loth to leave. She told Sobhdra about her family, giving her every detail.

The two women began to see more of one another. Whenever Sobhdra went to the river to bathe, she would take Saman along. Saman too felt she had to see Sobhdra at least once a day.

Saman was like the fish, that finding itself on the beach tosses in agony, but becomes exuberant when it is back in the water. In Sobhdra's company she felt as though she was in a heavenly river where she forgot her troubles and was happy.

Saman would help Sobhdra in whatever she was doing. Sometimes she cooked breakfast for Padam Singh. Sometimes she made his paan. In her opinion there was no woman in the world as gracious as Sobhdra and no man as noble as Padam Singh.

Once Sobhdra was down with fever. Saman was loth to move from her side. She'd go home for the shortest possible time, quickly cook some food and then run back. But Gujadhar was jealous. He did not trust Saman.

It was the month of Phagun. Saman was troubled about what clothes to wear for the Holi festival. Gujadhar had lost his job with Sethji a month ago, so now their steady income was limited to fifteen rupees. Saman had asked Gujadhar many times for a tanzeb* saree and a silk muslin jacket, but he always made some excuse. She thought, 'How can I go to Sobhdra's house for Holi in these old clothes?'

Meanwhile, Saman received news of her mother's death. Her sorrow was not as great as it should have been because Saman's feelings for her mother had become embittered. However her death provided the daughter with the excuse she needed for not being seen in fine new clothes on Holi. She told Sobhdra, 'I am orphaned, and have no more desire for clothes and jewellery. This sorrow has killed all such pleasures. No one knows what I have to bear.' She repeated this to her friends, and they praised her for her devotion to her mother.

One day Saman was sitting with Sobhdra, reading the Ramayan. Padam Singh came home in an ebullient mood and said to his wife, 'I've won.'

'Really?' an excited Sobhdra exclaimed.

Padam Singh: Is there room for doubt?

Sobhdra: Then give me my share. You've won your victory and I've won my bet.

Padam Singh: Yes, yes, you'll get your money. But be patient; my friends are insisting that I hold a banquet to celebrate.

Sobhdra: Well, that is quite a reasonable request. We'll have to do something.

Padam Singh: I had suggested an ordinary party, but it didn't satisfy my friends. They are insisting on a musical evening with Bholi Bai.

Sobhdra: Well why not? It would not cost you a fortune. And besides, Holi is coming. You could kill two birds with one stone.

Padam Singh: It is not the expense that is bothering me. It's the principle of it.

Sobhdra: Forget the principle this time.

Padam Singh: Das will kill me.

Sobhdra: Let him say what he likes. Not everybody will listen to him.

Pandit Padam Singh had just won the Municipality elections after many unsuccessful attempts. A feast was being prepared to celebrate this victory. Although he was a very principled man, Padam Singh did not have the strength of will to stand by his principles under pressure. It was partly politeness and partly his simple nature that were responsible for this. Another reason was his fear of the taunts of his friends. Babu Bithal Das was a close friend of Padam Singh. He always objected to dance performances which featured prostitutes. Moreover he had set up a reformist association to rid society of this base custom. Pandit Padam Singh was one of his few supporters, and therefore feared his condemnation. However, Sobhdra's urging

banished his hesitation and he complied with the demands of his more
convivial friends. It was decided that Bholi Bai would be asked to
perform. Four days later, on Holi, Padam Singh's drawing room had
been prepared for the dance. His friends were sitting on the handsome
carpets, and Bholi who was sitting with her musicians in the centre,
accompanied her melodious singing with lavish gestures. The room
was brilliant with electric light. Attar and rosewater were liberally
sprinkled. Pleasure and good taste ruled the evening.

Saman and Sobhdra were sitting on the balcony and watching the
proceedings from behind the blinds. To Sobhdra the songs gave no
pleasure. It surprised her that the audience was listening to them so
raptly. Saman had a finer ear for music, and a natural love for it.
However, one song left an impression on her. Bholi was singing:

> *Burn this Holi*
> *My love is far away*
> *I have no peace*
> *Burn this Holi*

Saman hummed the song slowly and was pleased at being able to
get the tune, though she could not render the words. But the song was
not the only thing that occupied her attention. She noticed that
hundreds of eyes were fixed on Bholi. There was such thirst, such
entreaty in those looks! The pupils brightened and shone at every
gesture of Bholi's. And if Bholi turned to an individual, he would go
into a trance; if she addressed someone, that person would look as
though he had discovered a treasure and others would cast envious
glances at him. In this gathering were present the best dressed, the
most handsome, the most learned, and the wealthiest individuals, but
all were enthralled by this woman and there was not one who would
hesitate to lay down his life at a sign from her!

Bholi's instrumentalists were applauding loudly and the whole
house was echoing with the song. But in the company there were
those who preferred to talk of other matters in whispered tones.

Said Munshi Abul Wafa, 'These people are quite powerless really.
All this is just pomp.'

Tegh Ali replied, 'He may have become a lawyer, but his father
was a nobody.'

On the other side of the room, two gentlemen were airing their
grievances: 'This is what they call justice. After working so hard,

when there is a chance of finding some reward, somebody who has influence to back him, bags it all.'

In one corner, Mr Kapatae was telling Mr Rudra, 'I am an outspoken person. Why should I fear anyone? They* appointed Mubarak Hussain, I appointed Gobind Ram. They fired Girja Sahai, I expelled Navazish Ali. They usurped rights. I did the same. This is the kind of treatment they can appreciate.'

Mr Rudra stated, 'Your presence has been very reassuring. Who knows what intrigues they would have contrived otherwise.'

Mr Kapatae said with pride, 'Just you watch, see how I shut him up. I'll go inspect the cattle house, and I'll definitely find a prey or two. Now look at this unfairness; The qazis of Qazibagh have not even paid up last year's dues not to mention the present year's, but nobody is making any demands from them. On the other hand, the Thakurs of Baghwar were served with summons because they failed to pay just one instalment this year. Wazir Ali's yearly income is not less than five thousand a year but no tax has been demanded from him. Meanwhile poor Gharib Das earns barely one thousand a year but his ledgers were rejected and he was made to pay five thousand rupees. Such injustices happen every day. So now I will adopt the same tactics.'

Mr Rudra said, 'Our people have no moral courage, while they give their life for theirs.'

Saman's thoughts were running on totally different lines. She asked herself, 'What magic does this woman possess? Is it the magic of beauty? but I am not bad looking either. She is dark; I am fair. She is stocky; I am slim.' There was a life-size mirror in Sobhdra's room. Saman stood before it and looked at herself attentively. The colours of Holi had tinged her pale skin with red, she compared her own looks with Bholi's. Her heart decided, 'I am much more beautiful than she is.' She then went to Sobhdra and asked her, 'Please don't mind my asking you this, but do you think this court beauty is prettier than I am?'

Sobhdra looked at her with surprise, then asked her, smiling, 'Why do you ask?' Hanging her head shyly, Saman said, 'For no particular reason; but do tell me.'

Sobhdra: She lives in luxury so her body is delicate; but your complexion is far better.

Then Saman began to think, 'So it must be on account of her adornments, her jewellery, and her dress that she infatuates people. If

I could adorn myself and wear jewellery and clothes like hers, wouldn't I look far more beautiful? But where can I find all that?'

'Is it her singing that bewitches people? But my voice is much better than hers. If somebody were to give me lessons for a month, I would be able to sing far better than she can. And as for the rest, I also know how to throw sidelong glances. I too can smile with my eyes lowered!'

For a long time Saman sat, comparing her own circumstances with Bholi's. Finally she came to the conclusion that Bholi was free while she herself was in chains. 'She keeps her shop open so there is a rush of customers. My shop is closed so there is nobody waiting. She is not afraid of the barking of dogs, I fear even whispers. She has discarded the purdah, I am behind the veil. She chirps away on treetops, while I toss and suffer in my cage. She has discarded diffidence and I am still tied to its apron strings. And it is diffidence that makes me fear for my reputation and that is what holds me in thrall to other people.'

It was past midnight. The function was over and people were going home. Saman also left to go home. It was dark all around her and in her heart there was the darkness of despair. Her legs were carrying her home, but very slowly. Pride has a distaste for penury, which was why her heart abhorred the house that was her home.

Gujadhar had returned home as usual at nine o'clock. The door of the house was locked. He couldn't think where she could have gone at this hour. A widowed seamstress lived in the neighbourhood. When he went to ask her, she informed him that Saman had gone to Sobhdra's on some business, and she gave him the key. Gujadhar came home and unlocked the door, cooked food, and sat by the door, waiting for Saman. By ten o'clock he was too angry to eat, so he threw the food out and locking the door from within, went to bed. He had decided that he would not unlock the door, no matter what. 'I'll see where she goes', he thought. But he couldn't sleep for a long time. At the slightest sound he would come to the door with a stick. His fury was such that had Saman arrived then, she would have paid heavily for her imprudence. After eleven o'clock however he succumbed to the ogre of sleep.

When Saman finally did arrive at the door of her house, she could hear the clock strike one. The sound echoed in her every nerve; so far she had been under the impression that it was no more than ten or eleven o'clock. She was petrified. She looked in through a crack in the door. The room was full of smoke from the lamp. Gujadhar was sleeping on his back, with the stick in his hand, and was snoring loudly. Saman was trembling. She did not have the courage to knock.

She thought, 'But where can I go at this hour? Sobhdra's door too would be locked now. They must be fast asleep. If I shout loudly enough they will open the door, but what would they think? No, it's best to stay here. It's already past one. In three or four hours it will be light.' And so thinking she sat down. But she was troubled with the thought that if someone saw her sitting there, she would be taken for a thief.

In the month of Phagun, cold night breezes blow. Saman's body was protected by just a torn silk tunic. The breezes pierced her bones like arrows. Her limbs shivered with the cold. And to make it all worse, the stink from the open drain was such that it was hard to breathe. Darkness, like a heavy cloud, was making its presence felt all around. Only rays of light from Bholi Bai's parlour were throwing furtive looks at the unlit street.

Saman was thinking, 'How unfortunate I am. On the one hand are those women, who are sleeping, comfortably pillowed, while their maidservants massage their limbs. On the other hand it's my lot to sit here, bewailing my fate. Why do I have to bear all this? Sleeping on a broken bed, in a hut, eating dry bread and putting up with rebukes every day. Why? Just for the sake of a good name. But what does the world think of such a sacrifice, what value does my self denial have in its eyes? Don't I know that! In the festival of Dasehra,* on the occasion of Muharram,* in Bini Gardens, in temples, everywhere I see her. I used to think that such women were loved only by the wicked, but today I know that they have equal access among the upright. The lawyer is such a good man. But how taken he was by Bholi!'

Thinking along these lines, Saman got up to knock at the door. 'I'll take whatever is coming' she thought, 'Why should I tolerate more hardship? After all, what am I doing it for? I would not be giving up a luxurious life. He has not provided me with one. I work hard all day to get no more than my daily bread. Yet I must put up with his bullying.' But one more look at the stick beside the sleeping Gujadhar took away all her courage. The frightened animal within her silenced the rational being.

Suddenly Saman saw two constables, with sticks resting across their shoulders, come towards her. In the dark they looked like ogres. Saman was petrified. There was no place to hide. She thought, 'If they see me sitting here they are bound to question me. What can I tell them?' She leaped up and knocked vigorously at the door, crying loudly, 'Open up. I've been shouting for the last two hours. Can't you hear?'

Gujadhar woke up with a start. He had had his first sleep. He got up and opened the door. There was some fear in Saman's voice that impelled him to do so. As soon as she came in, Saman said in an angry voice, 'How soundly you sleep! I've been shouting for two hours but you don't hear. My limbs are frozen with the cold.'

Seeing her before him, his anger returned. He said, 'Don't act haughty with me. Where were you all night?'

Saman spoke out boldly, 'What are you saying! "All night"? I went to Sobhdra's house at nine o'clock because she sent for me; returned at ten, and have been shouting at your door ever since. Or it may have been eleven o'clock. Does anything trouble you at all when you are asleep!'

Gujadhar: You came at ten o'clock?

Saman replied rashly, 'Yes. Yes, at ten o'clock.'

Gujadhar: That's a lie. I heard the clock strike twelve before I went to sleep.

Saman: Perhaps you did. You are never aware of anything but you were counting the hours!

Gujadhar: No more of your tricks. Tell me truthfully where you have been. I see you in your true colours today. I am not blind. I know well the cunning of women. Now tell me the truth or we'll settle everything today.

Saman: I've already told you that I came back at ten or eleven. Don't believe me if you like. Forget the expensive jewellery you were going to buy me!* Your sword is forever hanging out of its scabbard. Why are you so arrogant?

Even while she was saying this, Saman was startled by her own words. She knew that she had said too much. The decision she had reached while she sat outside by the door and all the arguments that led to that decision seemed to have receded in her memory. Custom, and the ideas that normally dominate our thinking, block out sudden revolutions in thought.

Gujadhar was stunned by Saman's bold speech. This was the first time that Saman had dared to speak to him in this manner. Infuriated, he said, 'Do you think you can do what you like, and I never say a word? Where were you all night? When I ask you this you say, "I don't care for you" and "What do you do for me?" It is the city that has changed you, and you are learning the ways of your friends. How often did I tell you not to associate with those witches, but you never heeded me. Unless you tell me now where you spent the night I will not let you stay here. And if you refuse, then you will have nothing more to do with me anymore.'

Saman replied timidly, 'I went to the lawyer's house, and nowhere else. If you don't believe me, go ask them. The singing lasted late into the night and Sobhdra wouldn't let me leave.'

Gujadhar said sarcastically, 'So now it's the lawyer that you fancy. Why then should you care for me?'

The taunt had the effect of a dagger plunged into her heart. A false allegation is intolerable. Saman spoke in a changed tone, 'What are you saying! Why give a bad name to a decent man. I was late, so do what you like. Beat me if you want to. But why drag in that poor man? He doesn't even set foot in his room as long as I am there.'

Gujadhar said, 'Don't try to hoodwink me, woman! I've seen many such "decent" men. If he is a god, go live in his house. This house is not good enough for you. You are getting too ambitious, you can't live here.'

Saman could see that matters were taking a serious turn. She would have taken back much of what she had said, if she could. But no arrow will return to the bow once it has been shot. So she wept, and said, 'May my eyes be torn out if I ever gazed at him. May my tongue fall out if I ever spoke to him. I go over to Sobhdra's to divert myself but if you forbid me, I'll stop.'

Once a suspicion has taken root it is difficult to banish it. Gujadhar thought, 'She has softened only to appease me.' He spoke even more harshly, 'No, do go. They will provide you recreation, and good food, and velvet mattresses to sleep on. There will be song, dance, and merrymaking, always.'

Taunts affect anger in much the same way as oil affects fire. Taunts crush the heart as a chisel shatters ice. Saman could not brook the false accusation. Shaking with fury, she said, 'Hold your tongue! You've said enough. For the past hour you've been haranguing me, just because I was patient. You are now accusing me of infidelity!'

Gujadhar: Yes, I believe you are deceitful.

Saman: You are accusing me falsely! God will be your judge.

This was the same Saman whom Gujadhar used to worship; a provocative glance from whom could send him into a frenzy. But love is another name for selfishness. Stung by her retort, he said, 'Stop cursing me, and get out!'

Saman: Why don't you just say that you don't want to keep me. Why the false accusation? You are not my master. I can work and fend for myself.

Gujadhar: Well, are you going or do you want to just stand here and abuse me?

A proud woman like Saman could not tolerate such indignity. The threat of being driven out of the house had hardened her dangerous resolve. She said in a decisive tone, 'All right, I am going.' But one step in the direction of the door, and she hung back, her decision wavering.

Gujadhar thought for a minute, then said, 'Take your clothes and jewellery. They are of no use here.'

These words finally blew out the dim light of hope. Saman was now sure that the doors of the house were closed for her. Weeping, she said, 'What will I do with them?'

But Gujadhar threw out the small box which contained her possessions, and the last thread of hope was broken. She picked up the box and left the house. Yet hope lingered, like life shuddering in a freshly decapitated body. She imagined that Gujadhar would come after her to console her and take her home. So she stood quietly on the road, facing the door, the corner of her garment wet with the tears she was wiping. Suddenly Gujadhar slammed shut both shutters in the doorway.

Those doors had been the doors of hope. It was pointless for Saman to stand there any longer. What effect could her tears have on Gujadhar now? She began to think, 'Where shall I go now?' Instead of repentance and regret she now felt only anger at Gujadhar. She had not, to her knowledge, done anything for which she deserved such a harsh punishment. It is true that she had returned home late. But for this, a rebuke would have been enough; such brutal treatment was unjustifiable. She had done everything in her power to appease Gujadhar. She begged him, and wept, but he not only insulted her but had made a false accusation as well. Now, even if he tried to placate her, Saman would not agree to go back. When he showed her the door he had told her, 'Go, and never show your face again.' These words had pierced Saman. 'Am I so unlucky that he does not want to see my face? Do all the women in the world have husbands? Aren't there any forsaken women? Well I too am forsaken now. I will work or beg to feed myself but I'll never show my face to him again!'

What a difference there is between the breezes of spring and the hot winds of summer. One is pleasant and life-giving, the other is poisoned and fiery. Love is the breeze of spring; hate is the hot wind of summer. The blossoms that it takes the spring breeze months to nurture, are withered by a single blast of summer's hot wind.

– 11 –

A short distance from Saman's house there was an uninhabited verandah. Saman lay down there, pillowed by her box. It was past three. She spent two hours in wondering where she should go. Among her acquaintances was Harya, a woman of easy virtue. She would have given her shelter, but Saman decided not to go there—she still had some self-respect left. She was now free, in a way, and could act upon the disastrous ideas that she had been entertaining for some time. Now nothing came in the way of her following a life of pleasure. Yet like a child who is delighted to see a cow or a goat at a distance but hides his face in fear as soon as he finds the animal near him, Saman, standing at the doorstep of her ambitions was unable to step in. A combination of diffidence, regret, and disgust had fettered her. She instead decided that she would go to Sobhdra's house, cook for her, serve her, and stay there. The rest could be left to the will of God.

She hid her box in her clothes and arrived at Pandit Padam Singh's. His clients were busy with their morning ablutions, some were meditating and wondering at the same time if their witness would change his story. Others were ostensibly telling beads but actually calculating on the beads what their expenses were likely to be. The cleaning man was collecting bread left over from the night before. Saman hesitated in going in, but when she saw Jitan the palanquin bearer coming, she quickly went in.

Sobhdra asked her with surprise, 'What brings you here so early today?'

Sorrowfully, Saman replied. 'I've been turned out of the house.'

Sobhdra: What! Why?

Saman: Because I went home late from your house yesterday.

Sobhdra: Just for that! What fuss over a little thing. He must be a strange man. Wait, I'll send for him.

Saman: No, don't call him. I wept bitterly but it had no effect on him. He grabbed my hand and showed me out. He thinks I owe him my sustenance. I'll shatter his arrogance!

Sobhdra: Don't say such things. I'll send for him.

Saman: I don't want to see his face.

Sobhdra: Are matters so bad between you?

Saman: Yes, they are. I have nothing more to do with him.

Sobhdra thought, 'She is still angry, but her anger will subside in a day or two.' She told Saman: 'Well go and have a wash. Your eyes show that you haven't slept all night. Sleep for a while, then we can talk.'

Saman: If I could afford to sleep in comfort I would not have had to live with a vile man like him. I am under your protection now. If you keep me I'll stay, if not, I'll go drown myself. Give me a corner in your house to live in and I'll serve you to the best of my ability.

When Pandit Padam Singh came home Sobhdra told him the whole story. Padam Singh was perturbed. It seemed to him that to allow an unknown woman to live in his house without the consent of her husband was not appropriate. As a lawyer, the legal aspect of the matter concerned him. He decided to send for Gujadhar and bring about a reconciliation between the husband and wife. 'It's best that the woman leaves this place' he thought and at once sent a man to fetch Gujadhar. But Gujadhar was not at home. After he came back from his chambers Panditji again sent for Gujadhar. This time again the mission was unsuccessful.

Meanwhile, as soon as Gujadhar found out that Saman was in Padam Singh's house, his suspicions were reinforced. Now he took to vilifying Padam Singh. First, he went to Bithal Das who thought Gujadhar's disclosures were no less than a divine revelation. This servant of the people and enemy of the scourges of civilization, was a strange mixture of generosity and narrow-mindedness. In his large heart there was compassion for the whole world, but there was no space in it for his opponent. Scandal is easy to believe. Hatred is convincing. Ever since Padam Singh had held the dance performance in his house, Bithal Das had held a grudge against him. He was overjoyed to hear Gujadhar's story. He immediately gave the news to Padam Singh's friends, colleagues, and relatives. He would say to people, 'Didn't I tell you that affair would have consequences? He has made a Brahmin woman leave her house and has lodged her in his own. The poor husband is going from pillar to post. He has gone quite berserk. This is the result of the standards set by high education, and the legacy of civilization. When I saw that woman frequent his house I straightaway smelled a rat.'

The strange thing was that even those who could be counted among Padam Singh's best friends, and knew him well, believed the story. We do not find it necessary to look for proof before believing slander.

The next day when Jitan went early to the market on some business, the news was already circulating. Shopkeepers asked him, 'Tell us Chaudhry, does the new bride come out before you or does she observe purdah?'

When Jitan came home after hearing all the jeers, he said to Padam Singh, 'Bhaiya, the mistress has lodged Gujadhar's wife in the house, and the people in the marketplace are insulting you. It seems she has come here because she has quarrelled with her husband.'

Gujadhar left no stone unturned in slandering Padam Singh. Before long, everyone who lived in the neighbourhood had heard the story. Groups of people collected in shops to discuss the matter. Calumny is readily believed. The grocer, a woman called Nazeeran, was sprightly and good looking. She used to go to Padam Singh's house to sell spinach. She swore that she wouldn't set foot in the house of 'that scoundrel' again. The old milkman who supplied milk to Padam Singh's household, had a young, squint eyed, sparse-haired daughter. He told his wife, 'Don't send the girl to their house. He is supposed to be a great man but see how he behaves!'

But in the same neighbourhood there also lived some affluent women who were more generous and who regarded human failings philosophically. They were unaffected by this turmoil. Zahooran was a big woman. Her neighbours used to call her 'cannon' or 'motor car.' She used to sell wood, dung cakes, cooking pots, and kerosene oil. Her shop used to be crowded with a certain type of women. They indulged in a particular kind of talk, day and night. There, Jiyori gave her verdict in Saman's favour, 'Which tree has never known the wind? The poor girl was married to that rustic. She lived in poverty. Now perhaps for some time she will be able to enjoy some comfort. After all, why else does God give good looks? What good is a pearl necklace hung round a pig's neck?'

When Padam Singh heard Jitan's account he was stunned. The world seemed to collapse around him. He was getting ready to go to work. One arm was in the sleeve of his garment and the other was out. But he was not aware of this. His worst fears—of attracting a scandal—

had been realized. Now he understood the reason for Gujadhar's negligence in acknowledging his messages. He understood also why he had become the focus of meaningful glances from all sides in his place of work. He understood why for the past two days people had been constantly honouring him with visits. He solved the mystery of why some informal souls had been talking to him in riddles. Silent and immobile, he stood thinking, 'What can I do except to turn her out? What she may have to suffer is not my business. Nor can I do anything about it. My name has to be cleared.' He was irritated with Sobhdra. He thought, 'Why did she have to take her in? She did not even consult me. She is safe sitting at home. It is I who has to face the world and am being humiliated. But if I turn her out where will the poor girl go? I don't think she has any other place of refuge here. And what will she think of me? Inhuman, heartless? Gujadhar will probably not let her return home. He has not bothered to enquire after her even though it is two days since she left. It shows he has decided to forsake her. But sending her away is the only way to protect my own reputation.' He said to Jitan, 'Why didn't you tell me about this earlier?'

Jitan: Sir, I only got to know of it today. Otherwise, would I have kept it from you?'

Padam Singh: All right. So, go in and tell Saman that her stay is causing a scandal. She must leave the house today whatever happens. But listen, say it tactfully, don't be blunt and hit her on the head with it!'

Jitan was very pleased. He felt for Saman the hostility that servants feel for those whom they perceive to be unimportant, yet favoured by their employers. He was an old man, and like others of his age found Saman's way of carrying herself, suspicious. He was illiterate, uncouth, and blunt. In spite of Padam Singh's exhortation, he went in and shouted loudly for Saman. Saman was making paan for Padam Singh. Hearing Jitan's loud voice she recoiled and stared at him timidly.

Jitan: Why are you staring? The lawyer has ordered you to leave the house immediately. You have no regard for his good name but he has to protect his reputation.

Sobhdra overheard this. She asked Jitan, 'What are you talking about, Jitan?'

Jitan: The master has ordered that she should go away immediately. People are gossiping.

Sobhdra: Go and ask him to come here.

Saman's eyes were full of tears. She stood up and said, 'No please don't call him. Nobody can force oneself into another's house. I will go away now and never come here again.'

Misfortune makes a person more sensitive. Lack of sympathy begins to seem like brutality, and reassurance, like a gesture of great kindness. Saman did not expect this from Padam Singh. With the self-centredness that comes with times of trouble she now believed him evil, egotistical, and callous.

She thought, 'Today you say you are afraid of losing your good name; your reputation is dear to you. But only yesterday you were sitting with a whore, looking very pleased, and enraptured by her. Your reputation was not at stake then, but today it is soiled!' Saman picked up her box calmly and casting a look of sorrow in the direction of Sobhdra, left the house.

– 12 –

By the time she reached the door, Saman was thinking, 'Where should I go?' The indignity she had suffered in this house wounded her more than Gujadhar's brutality. She realized now what a big mistake she had made in leaving her house. 'I took that step on the strength of my faith in Sobhdra. I used to think so highly of Pandit Padam Singh. But now I know that he is just a hypocrite. I have no home but my own. Why should I fawn on others when I have my own home. Hadn't I gone there in the first place with the certainty of spending all my life there? In a few days, when his anger subsides I could go back. How anger blinds us! I should never have come here in the first place.'

Thinking along these lines, Saman walked on. But after a while her thoughts took another turn. 'Where am I going! He will never let me step inside the house again. How humbly I had pleaded. But his heart did not melt. Only a few hours' absence made him so suspicious. And now I have been away for a full twenty-four hours, in the very house where I should never have set foot. To be sure he will drive me away. All I need is a place to live in. I can earn enough to feed myself, so why should I be dependent on anyone? What comforts did I enjoy in his house, for which it would be worthwhile giving up my freedom? If now he does agree to take me back, for fear of condemnation by the world, he will taunt me endlessly. Let me look for a house. Bholi Bai can help me find one. She used to ask me to her house again and again. Wouldn't she do this much for me?

'And what will happen if I go back to Amola? There is nobody there who I can call my own. Mother is dead. Shanta has a hard time living with them. Who will bother with me? Aunt will make my life difficult. She will stab me to death with her taunts. Let me ask Bholi's help in finding a house. Let's see what she has to say. If nothing else works, well there is always the holy river—the Ganges...'

Having finally determined what was the best course of action to follow, Saman walked to Bholi's house. On the way she looked hither and thither to make sure that she was not observed, especially by Gujadhar.

When she arrived at Bholi's door, Saman thought 'Why should I go in here? Perhaps some other neighbour could help me.' She was about to leave when Bholi saw her and motioned her to come upstairs. Saman obeyed.

When she saw Bholi's room Saman was amazed. She had been in this house once before but had seen nothing but the garden. The room was adorned with carpets, chandeliers, and pictures. On the carpet in the centre of the room was spread a cloth, elaborately embroidered with gold thread. In front of it was a life-size mirror, and in one corner was a low stool on which was kept a silver paandan. On another stool were placed a silver plate, a tumbler, and a khasdan.* Saman was astonished to see these elaborate arrangements. Padam Singh was a lawyer, yet his room did not have such adornments.

'Where are you coming from, box in hand?' Bholi wanted to know.

Saman: I'll tell you the whole story some other time. For the present please find me a house where I can live.

Bholi asked in surprise, 'But why? Have you quarrelled with your husband?'

Saman: No quarrel. But I can please myself.

Bholi: Just look at me. Yes, your face is betraying you. Tell me what happened.

Saman: Honestly, nothing happened. But why should I stay where I am an inconvenience?

Bholi: Why don't you tell me frankly why he is angry?

Saman: It was nothing to be angry about. But if he is angry then all is over.

Bholi: You can try your best to keep your secret, but I know what it is; and I'll tell you if you allow me. I've known all along that you two would never get along. Who ever heard of an Arab stead and a mule harnessed to a carriage at the same time. You should have been the mistress of a great house. But instead you were tied to that worthless fellow. Only you could have put up with him. Another woman would have left him long ago. If God had given me the good looks that He has bestowed on you, I would have piled up enough gold to build a wall! I don't understand you. Perhaps you have not been educated well.

Saman: I was tutored by a Christian lady for two years.

Bholi: Had she instructed you for two or three years more you would have understood the purpose of life—how we should enjoy it. We are not cattle for our parents to give away to whom they like, and thereafter become the property of that person! If God had intended you for adversity, why would He have given you such beauty? These vulgar customs thrive only in our culture where women are degraded. In other countries women are free. They can marry who they like, and if they don't get along, they can leave their husbands. But we never think of changing the age-old customs.

Saman replied thoughtfully, 'What can I do? One fears what people say, otherwise who would not want to live in comfort.'

Bholi: This is all the result of ignorance. My parents had married me to an old man. He was wealthy and I lived in comfort, but I couldn't stand him. I endured it for six months but then I left. Life is a gift, it was not given us to be spent in misery. What good is it if one cannot taste its pleasures? At first I too feared disgrace. But as soon as I left home, things changed. I became popular, and was fawned upon by people who mattered. I had been taking singing lessons before I left, I took some more. Very soon the whole city got to hear of me. And now there is no magnate, maulvi, pandit, or officer who doesn't consider it an honour to serve me. I sing in temples and mausoleums. People ply me with invitations. How can I believe I am disreputable? Just to divert myself if I send for him, the Mahant of your Krishan Temple would come running. Let whoever thinks so, call my success a disgrace!

Saman: How long would it take me to learn to sing?

Bholi: You need no more than six months. Nobody here cares for real singing. There is no need for highbrow classical music. Only popular love poetry and light classical music with accompanying gestures is enough to put you on the map. All they really want is good looks and engaging speech. God has given you both these gifts in abundance. I swear Saman that if once you break that iron chain of yours you will find that people run after you like madmen.

Saman said in a worried tone, 'But what I don't like is...'

Bholi: Yes, yes, I know what you are thinking. One has to be shameless with all kinds. At first it bothered me too. Then I realized that it didn't happen that way. The riff raff don't dare come here. Only those with money safely tucked away do. And they alone are worth ensnaring and keeping trapped. If they are the decent sort and one finds them likeable, then shamelessness is not an issue. But if you don't like them you should keep them dangling, and entangled in mere talk. Plunder and ravage them as much as you can, and they will soon drop out in frustration. Others will take their place.

It is true that at first one is shy, but isn't that so with one's husband too, in the beginning? Gradually there is no more shyness with the husband, and so it is with the others.

Saman said with a smile, 'Well, first find me a house.'

Bholi understood that the fish was nibbling at the bait, it was time to stiffen the line. She said, 'This house is at your service. Live here.'

Saman: I will not live with you.

Bholi: Why, will that give you a bad name?

Saman (embarrassed): No, it's not that.

Bholi: Would you be bringing disgrace to your family?

Saman: You are making fun of me.

Bholi: What other reason can you possibly have? Will Pandit Gujadhar Parshad be offended?

Saman: What can I say?

Saman had no means of refuting Bholi's arguments, as Bholi had already weakened all her objections by ridiculing them. Yet the aversion that one feels naturally for shamelessness and prostitution was throwing her resolve into disarray. She could not put her feelings and thoughts into words. Her state was that of one who looks greedily at ripe fruit in a garden, but cannot bring himself to pluck it even in the absence of the gardener.

Bholi interrupted her thoughts with, 'How much rent can you pay for a house, so I can instruct my maid to find one accordingly.'

Saman: About two or three rupees.

Bholi: And what will you do for a living?

Saman: I can sew.

Bholi: And will you live alone?

Saman: Yes, who can I live with?

Bholi: Now you are talking like a child. Silly girl, you have eyes and yet you are blind! You can't live alone in a house for one day. You are bound to be raped. It would be better for you to go back to your husband.

Saman: I'd rather not see his face. I won't keep this from you; only the other day when you were performing at the lawyer's house, his wife who is very fond of me, sent for me. She didn't let me leave until the end of the performance, which was around twelve or one o'clock. I got home late, and just for that he was so angry that he abused me endlessly. I swear on God that I tried my best to appease him. I wept, fell at his feet, but to no avail. He drove me out of the house. If somebody doesn't want to keep you in his house, you can't force him to. I went to the lawyer's, thinking I'd stay there for a few days and then see what I could do, but that brute maligned him. The result was that the lawyer sent word to me that I should leave.

You know, one put up with hardship, but at least there was the satisfaction that one lived a respectable life. But now there is this dishonour to be borne as well. No matter what happens to me now, I will never step inside that house again.

Saying this Saman's eyes filled with tears. Bholi consoled her, saying, 'Go and have a wash first, and then eat something. Then we'll discuss your plans. I think you have not slept all night?'

Saman: Can I get some water here?

Bholi said with a smile, 'We'll arrange everything. My kahar* is a Hindu. So many Hindu gentlemen come here that I have employed a kahar for them.'

Bholi's old maid took Saman to the bathroom. Saman washed herself with soap. Then the maid dressed her hair and brought her a new silk saree to wear. When she was ready and went upstairs, Bholi looked at her with an envious smile and said, 'Go look at yourself in the mirror.'

When Saman stood in front of the mirror, she felt that she saw the goddess of beauty before her. She had never thought of herself as so beautiful. Her face lit up with pride and her eyes were intoxicated. She reclined on a couch.

Bholi said to her maid, 'Well, Zahooran, don't you think we can trap Sethji now?'

Zahooran replied, 'He will massage the soles of her feet, he'll be so perfectly docile.'

In a little while the kahar brought sweetmeats. Saman ate, had a paan, and then went and stood again before the mirror. She thought to herself, 'Why should I renounce this luxury to go back into that dark cage?'

Bholi asked her, 'What should I say if Gujadhar Parshad asks me about you?'

'Tell him I'm not here.' Saman replied.

Bholi had achieved her purpose. She was certain that Seth Balbahadar Das who had been avoiding her, would fall victim to this beguiling instrument of seductiveness.

Saman's present state was like that of a greedy doctor who when he goes to see a patient, who is also a friend, refuses to take the fee out of embarrassment but when it is dropped into his pocket, goes home smiling with pleasure.

– 13 –

Padam Singh had an older brother called Madan Singh who looked after the ancestral home. He was a petty landlord, and owned a small farm. He had just one son, whose name was Sadan Singh. His three daughters were, Munni, Chhanni, and Chunni. The wife's name was Bhama. An only son is a fortunate being. He gets plenty of sweet things to eat; and no bitter admonitions. As a child Sadan was stubborn and insolent. As he grew up he became irresponsible, bad tempered, and lazy. The parents did not mind; they had him, and it did not matter how spoilt he was. They could not bear to be parted from him even for a day. Padam Singh had repeatedly urged that the boy be allowed to go with him to the city where he would be admitted into a school, but the parents would not agree. The mother said, when there was plenty for him to eat at home, why should he be sent away. He would stay unlettered but at least he'd stay before their eyes. Sadan had read Urdu and Hindi in his village school. In Bhama's opinion any further education was unnecessary.

Sadan was always ready to go with his uncle. His polished manner, his soap and towels, shoes and slippers, watch and collar, all excited him. His own house was well-stocked but not with such stylish things. He would have loved to dress like his uncle and go out in a carriage. He was very respectful to Padam Singh and always heeded what he said. Although he was often rude and inattentive to his parents, he was affability itself when his uncle was present. His uncle's suavity and well-dressed appearance had won his heart. Whenever Padam Singh came home he used to bring him good clothes and shoes. Sadan accepted these presents with avidity.

Padam Singh always went to his ancestral home for the Holi festival. This time too, he had written a week before to say that he would come. Sadan was dreaming of silk achkans and new shoes. The day before Holi, Madan Singh dispatched the palanquin to the station, both in the morning and in the evening. The next day too the palanquin was sent at both times, only to return without the passenger. At this very time, at Padam Singh's, preparations were being made for Bholi Bai's concert. How could he go? Back in his village this was the first Holi on which he was absent. Bhama wept and Sadan was in despair. No new clothes! How could he take part in the Holi festivities? Madan Singh too was dejected. An air of depression hung over the whole household. The women from the village who had come to play Holi, consoled the unhappy Bhama. They said, 'How can somebody who is not of your own flesh and blood ever be yours?' The husband and wife must be having a good time in the city. Why should they care for the village?' There was singing and dancing. But Bhama did not enjoy it. Madan Singh used to drink a lot of bhang on Holi, but he did not touch the drink this time. Sadan hung around with a long face, scantily clothed. In the evening he told his mother, 'I am going to Uncle's.'

Bhama: Who cares for you there?

Sadan: Uncle is there.

Bhama: He is not the same person anymore. He has made some money. He does not care for you.

Sadan: I will go there.

Bhama: I don't want you to go there. Don't pester me.

The more Bhama tried to stop him the more obstinate Sadan became. Irritated, Bhama got up and went away. Sadan too went out. Obstinacy cannot be overcome by a frontal attack. The charge should be aimed at the flank. A startled horse will run if it is frightened. Bribed with grain, and caressed, it can be controlled. Sadan had secretly decided to run away to his uncle. 'Why should I stay here? These people are not likely to give me a silk achkan. At best they will have a kurta stitched for me. Just because they gave me the mohan mala they think they have done something splendid. They show it off to the whole village. Yes, I will go and nobody can stop me.'

After he had made up his mind, he waited for the right moment. When everybody had gone to bed he quietly left the house. The railway station was less than three miles away. The moon, still a crescent, had set. Beyond the village there was a bamboo cabin. When Sadan reached the cabin he could hear strange sounds coming from it. At first he was terrified, but immediately after, he realized that the sound was made by the swaying bamboos when they rubbed against one another. A little further down the road there was a mango tree. Many days ago a peasant's son had fallen from it, and died. When Sadan reached the tree he thought that there was somebody standing by it. His hair stood on end and he felt dizzy. But gathering up his courage he looked hard at the figure and finding nothing there he bounded forward and resumed his journey. Soon he had left the village behind.

Two miles beyond the village was a pipal tree. It was rumoured that it was a demons' den.

All the demons lived on the tree. Their leader was a demon who kept himself covered in a blanket. He would appear before travellers in his black blanket and wooden clogs, begging with his hands apart. When the travellers stretched out their arms to give him money he would vanish. Nobody knew the purpose of this mischievous behaviour. As a result few people came to the spot alone at night, and anyone who had the courage to do so was sure to relate some supernatural incident on his return. One would say, 'There was singing there.' Another had witnessed a council of demon elders. This was the last frightening experience that Sadan had to walk past, and he was building up his courage for it. But as he neared the spot his courage began to melt like ice. When a distance of no more than a furlong remained between him and the tree, his feet refused to carry him further, and he sat down on the ground wondering what he should do. He looked all around but could see no trace of a living being. Had he seen even one animal, he would have felt more confident. For half an hour he waited, hoping to see another traveller. But village roads are deserted at night. He thought, 'How long can I sit here? The train leaves at one o'clock. If I am late the whole exercise will have become futile.' So he gathered up his courage once again, and singing aloud hymns from the Ramayan, he resumed his journey.

He was trying hard not to think. But on such occasions thoughts, like summer flies, are persistent—shoo them off and they come back again. Finally he could see the tree in front of him. Sadan looked at it closely. It was late at night and the darkness had deepened. He could not see anything. He began to sing even louder. Every pore on his body seemed alive and alert. He looked here and there. He could see many different creatures which however disappeared as soon as he looked at them more attentively. Suddenly he felt that there was a monkey sitting on his right. He trembled with fear, but in a moment the monkey had become a heap of dirt. When he arrived under the tree he began to feel suffocated, and he could not utter a word. Now it was no longer necessary to beguile his thoughts because a concentration of all his attention, his senses, and his daring was required. His ankles were trembling and his chest was choking the organs within. Suddenly he saw something running. He sprang up. When he looked again he saw that it was a dog. He had heard that demons sometimes appear in the guise of dogs. His senses seemed to desert him and he stopped and stood silently as though waiting for the enemy to attack. Meanwhile the dog, with his head bent, walked away, giving him a wide berth. Extreme fear turns into daring. Sadan shouted loudly at the dog, 'Shoo.' The poor dog ran, his tail between his legs. Sadan ran after him for a short distance and became certain that it was indeed a dog. Had it been a demon it would definitely have shown a trick or two. His fear diminished but he didn't resume his journey immediately. When a penniless man finds a little money he begins to strut about, and so it was with Sadan. To regain his self respect he stood under the pipal tree for many minutes. Not only that, he even walked around the tree and tried to shake it with both hands. This was an uncommon degree of valour! Though now he seemed hard as stone on the outside, inside there was no stone—just trembling liquid. A little sound, a little movement, could very well have made the difference between life and death for him.

Having come out of this trial, Sadan walked towards the railway station, his head held high with pride.

– 14 –

After Saman left, Padam Singh was overcome with remorse. 'I should not have done what I did. God knows where the poor girl went. If she has gone home, that would be the best thing; but it is doubtful if she has. I hope she has not fallen into the wrong hands. A man at the end of his tether becomes shameless. For a young and beautiful woman to leave home is like words leaving the mouth. It was foolish of me. This is no time for me to stand on my dignity. She is drowning and should be rescued. People will talk, but no matter.' Thinking in this vein, Padam Singh began to get dressed to go to

Gujadhar's house. He was ready and had left his house but was uneasy at the thought that somebody might see him at Gujadhar's door, and what would Gujadhar himself think? What if he started a quarrel? Padam Singh went back to his house and took off his outdoor clothes.

Our extraordinary actions do not immediately follow the decision to act. We remain perplexed and indecisive to the last. When at ten o'clock he went in to eat, Sobhdra asked Padam Singh in an angry voice, 'Why did you have to persecute Saman early this morning? If you had to send her away you could have done so more tactfully. You sent that old Jitan with your message and he put it in the most unfeeling words. Poor Saman said nothing at all, just got up and left. I could not lift my head with embarrassment! Instead, if you had told me, I would have explained the situation to her. She is not a fool, she would have made arrangements and then left. Things need not have been done so crudely. You just upped and sent her marching orders. So afraid of a scandal! What if she hasn't gone home, will there be less of a scandal? Who knows where she will go; and who will get the blame for that?'

Sobhdra had been full of rage. Her rage boiled over. Padam Singh was listening with his head bent, like a criminal who has pleaded guilty. He had heard his own thoughts from Sobhdra's lips. He couldn't lift his head from embarrassment. He ate his food and went to the courts. It was the third day after the function. He had always been seen by his colleagues as a principled man, and they respected him. But for the last two days things had changed. The other lawyers would come and sit with him and confide in him, 'Sharmaji', they would say, 'a great singer has just arrived from Lucknow. Won't you arrange a concert?' Another would say, 'Sharmaji, have you heard the latest? Seth Chaman Lal has become infatuated by your Bholi.' Yet another would suggest 'Tomorrow is the day of ceremonial bathing in the Ganges. There will be festivities by the riverside. Why not throw a party and invite Sarasvati. Her voice is nothing much, but what a beauty she is!' Padam Singh was disgusted by such talk. He would think 'Why do they say such things to me. Do they think I am an agent for the courtesans? 'There was a marked change in the attitude of the court officials too. They would come and sit with him whenever they found the time, and smoking their cigarettes, would talk in the same vein.

Tired of these unwelcome visitors, Padam Singh would go and hide under some tree in order to avoid them. He would curse the hour when the function took place in his house. Today too he could not stay long at the court. Hearing the disgusting talk, he left early and arrived home at two o'clock. As soon as he reached his door, Sadan came up and kissed his feet.*

Padam Singh exclaimed in surprise, 'Sadan! When did you come?'

Sadan: By the train that's just arrived.

Padam Singh: Is everybody well, at home?

Sadan: Yes Uncle, they are all well.

Padam Singh: I couldn't go for Holi this time. Did Bhabi* mind?

Sadan: We waited two days for you. Then I became restless so I came.

Padam Singh: Does that mean that you came without their knowledge?

Sadan: I did ask to be allowed to come, but you know them; Mother wouldn't agree.

Padam Singh: Then they must be worried. You should have at least got somebody to accompany you. Anyway, I am glad you've come. I wanted to see you. Now that you've come, why not enroll in a school?

Sadan: That is why I am here.

Padam Singh sent a telegram to his brother which read, 'Don't worry about Sadan. He is here and will be enrolling in a school.'

After sending the telegram Padam Singh began to talk to Sadan about their village. The conversation continued until the evening. There was no tenant farmer, potter, blacksmith, or cobbler, about whom he did not enquire. Village life creates a close fraternal bond among the inhabitants of a village, which is absent from the lives of city dwellers. All village folks, great or small, are knitted together in this way.

That evening Padam Singh went out for some recreation with his nephew, though not in the direction of Bini Gardens or Queens Park. Instead, he headed towards Darga Kand and Kalanji's dharam shala.* His heart was uneasy and his eyes were anxiously searching all around for Saman. He had decided that if he found her he would not let her go again, no matter what rumours he had to put up with. If her husband claimed her she could go back to him if she pleased. Then he thought, 'Let me go to Gujadhar. It is possible that she has gone home.' Immediately he turned back to go to his own house. Many clients were waiting for him. He looked at their documents but his thoughts were elsewhere. As soon as he had got rid of them he left for Gujadhar's house. All the way, he kept looking around to make sure that he was not observed or followed, and at the same time he affected a nonchalant demeanour. When he reached Gujadhar's doorstep,

Gujadhar himself had just returned from work. That afternoon Gujadhar had heard the news that Padam Singh had turned Saman out of his house. However he suspected that this was just a pretence to hide the girl. But when he saw him standing before him he could not help standing up in deference. This is the due of status and rank. He got up from the string bed and made his obeisance. Padam Singh enquired dispassionately, 'Well Panditjee, your lady is back, isn't she?'

Gujadhar's suspicions were slightly allayed. He said, 'No sir, she isn't. Since she left your house I know nothing of her whereabouts.'

Padam Singh: You didn't look for her! What is it that has made you so angry?

Gujadhar: Sir, my anger was just an excuse. She herself wanted to leave. The madams in the neighbourhood had put her up to it. In recent months there was a change in her. On Holi she came home at one in the night. My suspicions were roused and I scolded her. That was all. She left the house for that.

Padam Singh: But if you had wanted you could have taken her home from my house. Instead you chose to dishonour me. Why should I have put up with disgrace? So today, early in the morning I sent her away. After all, we all care for our reputation. In this matter my only mistake was that on the day of Holi she stayed late in my house. If I had anticipated the consequences of that function, I would never have held it. I would not have let her come to my house. For this lapse you disgraced me throughout the city.

Gujadhar began to weep. He became certain that his suspicions had been false. Between sobs he said, 'Sir, you can punish me as you will. I am an ignorant man who believes whatever he is told. The Bank house gentleman, what's his name, yes, Bithal Das, he's the one who misled me. A day before Holi he came to my masters' shop where I work, to buy some fabrics. He took me aside and talked to me about you...now what can I say. It infuriated me, because I thought it came from an honourable man. He talks to the whole town about benevolence and humanity. One tends to believe such a do-gooder. I can't say why he feels such enmity towards you. He has certainly destroyed my home.'

Padam Singh felt as though somebody had branded him with a red hot iron. His forehead was covered with perspiration. He could take the thrust of a sword as long as it came as a frontal attack. But coming from the back even the prick of a needle was intolerable. Bithal Das was his confidant, a friend who had studied with him. Padam Singh looked up to him. Even when he did not agree with him he respected his good intentions, though his intentions were frequently not practical. If such a person could deliberately indulge in gossip, it could only mean that he was devoid of decency and bent on mischief. Padam Singh understood that Bithal Das had spread the rumour merely to defame him, that he was slandering him in order to degrade him in the eyes of the world. Furious and impatient, he asked, 'Would you say this to his face?'

Gujadhar: Yes, why should the truth not be told? Let's go. I will say it to his face and he will not dare to deny it.

In the heat of anger Padam Singh had agreed to go, but in a while the storm within him abated somewhat and he recovered his good sense. It occurred to him that going there now would complicate matters further, so he said to Gujadhar, 'All right, come when I call you. But meanwhile don't be negligent. Keep looking for your wife and if need be, I'll pay your expenses.' So saying Padam Singh went home. The blow from Bithal Das's hidden sword had wounded him grievously. He felt certain that Bithal Das had started this trouble out of sheer malice. It did not occur to him that Bithal Das had said whatever he did from good intentions and was himself convinced of it.

The next day Padam Singh took Sadan with him to enroll him in a school. But wherever he went he was told that there was no vacancy. There were twelve schools in the city, but there was no place for Sadan in any of them. Finally he decided to teach him himself.

All morning Padam Singh used to be busy with his clients. So he would teach Sadan after he returned home from the courts. However he could not get used to this new routine and had to give it up within a week. It had been his practice to read the newspaper or play the harmonium after he got home from work. Instead of these congenial activities, teaching Sadan necessitated browbeating his grown up nephew into working at his studies. In the process he would lose his patience, because it appeared to him that Sadan was excessively stupid and dull. If Sadan asked him the meaning of a word, he would look for a long time to find the word in the book where it first appeared. Then he would show it to Sadan and ply him with questions to elicit the meaning of the word. In all this, a lot of time was spent but very little was achieved. Sadan was afraid to open the book in his uncle's presence and had even started regretting that he had come at all. He thought, 'I wish I had stayed back in the village. He teaches me a few lines of the lesson but is angry for hours.' By the time the lesson came to an end Padam Singh would be so upset that he had no desire for recreation. He realized therefore that he had no aptitude for teaching.

In the neighbourhood there lived a teacher who agreed to give lessons for twenty rupees a month. Now the question was, how could the money be raised? Padam Singh was a man of fashion and his expenses were many. The burden of expense however did not mitigate his taste for elegance. He sat and thought about it for a long time but could think of no solution. Fashion is a romantic dream which feeds on the heart's blood, but its reward is no more than some applause. Finally Padam Singh went to Sobhdra and said, 'The teacher is ready to teach for twenty rupees.'

Sobhdra: Can't some other teacher be found? There are hundreds of teachers, but where is the money?

Padam Singh: God will provide the money.

Sobhdra: I have noticed that for these many years God has not favoured us very much. He has given us just enough to fill our stomachs. Is He now going to change his ways?

Padam Singh: No, but Faith bears fruit.

Sobhdra: When Faith bears fruit, usually debts bear fruit as well.

Padam Singh: You are only taunting me. Think of a solution.

Sobhdra: All I can say is, don't bestow your usual largesse on me.

Padam Singh: Now you are annoyed.

Sobhdra: What you are saying is annoying. You are well aware of our income and expenses. Where can I save? No rivers of milk flow from your house. There is no over-abundance of fancy food here. We can't do without the kahar. It is necessary to keep the cook. Where else can I cut costs?

Padam Singh: (abashed) You could stop the milk.

Sobhdra: Yes, stop the milk. But even if you don't drink it, Sadan must have it.

Padam Singh felt he was drowning in a river of anxiety. The cost of paan and tobacco was no less than ten rupees a month. There were also some other small expenses that could be curtailed. But even to mention these would mean drawing down Sobhdra's wrath. It became apparent to him that in this matter Sobhdra's sympathies were not with him. Therefore, he began to take stock of the expenses in the men's quarters, and this gave him some hope. He said, 'The amount spent on lights and fans can be reduced somewhat.'

Sobhdra: Certainly. It's not necessary to have lights. You could go to bed before dark. If there is a visitor he'll go away after he has shouted himself hoarse, to no avail. Or you could go out for recreation and not return until nine or ten. As for the fans, you can always use the hand-held types. After all, there was a time when there was no electricity; yet you managed to keep your senses.

It seemed that Padam Singh had sworn not to be embarrassed. He said, 'So, shall I cut down on the horse's rations?'

Sobhdra: You are generous. Why does the horse need his rations when he can live on grass? All that might happen is that he'd be emaciated; but he can still take you to the courts. Nobody will be able to say 'The lawyer does not have a conveyance.'

Padam Singh stood his ground manfully in the face of this dart of derision. He said, 'I contribute two rupees monthly to the girls' elementary school, two rupees a month go towards paying club bills, and one rupee a month is donated to the orphanage. Shall I stop these expenses?'

Sobhdra: Why not? After all charity begins at home.

This time too Padam Singh showed fortitude. He said, 'I could cut my expenses here and there and manage to save about fifteen rupees a month. You will have to bear the burden of providing the rest of the sum. I have never asked you for an account of your expenditure. But find this amount somehow.'

Sobhdra: It's no problem at all. It will be done. Starting from tomorrow one meal a day, only, will be served. Why must we cook two meals in one day, anyway? After all there are millions of people in the world who eat just once a day, and who are neither weak nor ill.'

Padam Singh had come to the end of his patience. So far he had been conducting himself with the same maturity and fortitude that helped him to ignore hostile witnesses. But domestic squabbling unnerved him, and though he usually restrained himself, this last blow was more than he could endure. He said, 'So you don't want me to hire a tutor for Sadan, and would let him waste his life? Instead of understanding my problem you ridicule it. Sadan is the son of the brother who put me in school though he himself had to vend rice and lentils, carrying them in bundles on his head. I have not forgotten those days. When I remember his hard work I want to fall at his feet and sob for hours. You find it painful to cut expenditure on electricity, paan and tobacco, upkeep of the horse, but Bhaiya bought me good shoes while he himself went barefoot. I wore silken clothes while he made do with torn garments. His kindness and goodness have placed such a burden of obligation on my shoulders that I cannot lighten it in a lifetime. For Sadan I am ready to put up with any adversity. For his sake I am willing to go to the courts on foot, starve, and clean his boots with my own hands. And if I were reluctant to do so I would be the most disloyal and ungrateful creature in the world!'

Sobhdra's expression wilted with remorse. Although Padam Singh had meant what he said, sincerely, she thought that he had said it to embarrass her. She was even more mortified because she knew that her feelings on this matter had become evident to him. It was true that she had been displeased by Sadan's arrival, and also that she considered it foolish to incur so much expenditure on his account. Bending her head she responded to his chiding, 'When did I say that a tutor should not be employed for Sadan? By all means, do whatever needs to be done. Since your brother made such sacrifices for you, it is only right that you should spare no expense for Sadan. I will do whatever you ask me to. You had never brought up this matter before, so I had the impression that it wasn't necessary to spend so much money. There was no need for this procrastination; by now he would have acquired some education. Since you have taken the decision to educate him so late in his life, then not a day should be wasted.'

Sobhdra had taken her revenge for the mortification that she had been caused. Padam Singh had to acknowledge his mistake. He realized that he had not been as intensely conscious of his obligation as he claimed. 'Had Sadan been my own child his education would have been my first concern. It is I who has been guilty of delay in the discharge of my duties. In her reluctance, Sobhdra has merely taken the cue from me,' he thought. He made no reply to Sobhdra.

Sobhdra regretted her response. She made a paan and offered it to him—a peace offering. Padam Singh took the paan—the offering had been accepted and peace was re-established.

When he was leaving, Sobhdra asked him, 'Have you had news of Saman?'

Padam Singh: No, she has vanished. Gujadhar has disappeared too. They say he has left his house and gone away.

The next day Sadan started his lessons with the tutor. The teacher would teach him and leave by nine o'clock. After that Sadan would eat, and then go to bed. He had neither friend nor companion. There was no recreation. Sadan was lonely and bored. He worked out in his room early in the morning, and that was his only diversion. In his village he had had a small arena built for himself, but this facility was not available in the city. In the evening Padam Singh had the phaeton prepared for him. He himself was fond of such outdoor recreation and used to take the direction of the park or the cantonment. But Sadan never headed that way. He had no taste for fresh air and the quiet but thought-provoking beauty of nature. Appreciation for such things requires a certain refinement and sensitivity—qualities that were absent from his character.

Sadan was in the prime of life, a time when vanity is at its height. He was a tall and good-looking young man. All his life he had lived in the country where he never learned to read or write. He had never been in awe of a teacher nor had he experienced fear of examinations. He was in the habit of drinking immense quantities of milk and devouring large amounts of ghee. To cap it all he was in the habit of taking plenty of exercise. He had therefore developed a well-proportioned body, a broad chest, and a strong neck. However his face lacked the gravity and refinement of expression which is the mark of an educated and cultivated mind. His countenance was manly, rough, and hard. He was not a grafted garden plant, but rather, a sturdy tree from the wild. In his gaze was an alluring absorption, and in his gait there was a proud intoxication. But youth is not miserly; it does not hide its treasures. It is a madness that is happy to live among the starving. In the remoteness of a park or a meadow, who could have set eyes on him, or applauded his radiant good looks? He therefore liked to wend his way in the direction of the marketplace or the red-light area. Here all eyes would look with admiration at his manly beauty. Young men looked at it with envy, old men gazed with approval, at the same time regretting that this graceful youth must soon be scorched by the flaming desert of life.

But on the decorated upper storeys of the shops, the sight of this ravishing young man would cause frenzy among the inmates. The courtesans would stand on their balconies. Hundreds of eyes, filled with coquetry and eager invitation would be gazing down at him. The atmosphere would be charged with excitement and enticement, and expectations would rise high as they watched him. Where would this exotic, lost pigeon alight? Whose snare would entrap this golden bird? Among the enchantresses there were many who were themselves vulnerable to the enchantment of beauty, and it was they who were yearning for the pleasure of intimacy with him. His air proclaimed that he was waiting with bared chest for the arrows of love. His look of affliction was proof that he desired such wounds. This was the fascination that attracted these seasoned seductresses to him.

Sadan lacked the serenity that ensures chastity. Nor had he the self restraint and gravity which are vouchsafed by self esteem, and which guard against brazenness. Moreover, he had not yet learnt to dissemble. In the marketplace his phaeton would move slowly, and his gaze would fix on the upper storey parlours of courtesans. In our youth we are proud of our weaknesses, and after youth has passed, we take pleasure in the display of our good moral qualities. Sadan wished to appear as the consummate lover, though more than the love, he longed for the notoriety of passion. If at the time he had had access to a friend in whom he could confide, he would have recounted to him a concocted tale of debaucheries. Had somebody accused him of immorality he would have been proud, not abashed.

Sadan had not yet learnt to be selective, so all the women of the street appeared to him equally valuable. Each one of the cups proffered in open invitation, held for him the same charm. Attracted only by the light, the moth does not know whether it is gas, electricity, or kerosene which sustains it. Finally, it got to the stage where Sadan could think of nothing but the courtesans' district. The looks, gestures, and smiles of the women stayed in his thoughts night and day. He began to detest the sessions with his tutor and would feel his burden lighten only when he left. For the rest of the day he would sit before the mirror or brush his suit. As soon as it was evening he would dress smartly and go off in the direction of the marketplace or the red-light area. Gradually he became bolder and the desire to render his feelings into deeds, grew. But the problem was that two men always accompanied him on the phaeton like keepers of his conscience. In their presence he did not dare to touch the blossoms in the garden of pleasure. He began to think of a way to get rid of the two men. Finally he thought of a scheme.

One day he said to Padam Singh, 'Uncle please give me a good horse. In the phaeton I have to sit like a cripple and it is no fun. Riding a horse will give me some exercise and also I'll get a chance to improve my horsemanship.'

From the day Saman left, Padam Singh had been depressed. The clients complained that he had changed. He had become irritable. They grumbled, 'If he does not listen to us how can he argue on our behalf? And if we are paying fees we can find another lawyer. Why should we retain him when he spends so much time wandering about in the lanes.' And so Padam Singh was losing his clientele by the day. The resulting reduction in his income further dampened his spirits. When he heard Sadan's request he was troubled. 'Why don't you ride the cart horse? He'll get used to the bridle in a few days.'

Sadan: No, he will not be able to stand so much travelling. He has no pace and he can't gallop. He is exhausted when he returns from the courts, how can he put up with more work?'

Padam Singh: All right, I'll see what I can do. If I find a good animal I'll buy it.

Padam Singh had tried to evade the issue. Even an ordinary horse would not cost less than about three hundred rupees. And a further expense of about twenty-five rupees a month would have to be incurred for its keep. He could simply not afford such an outlay. But Sadan would not be appeased. He repeated his demands every day, and then several times a day. Padam Singh would grow cold at his approach. Had he explained to him his financial predicament, Sadan would have been silenced. But he was not willing to burden him with his own troubles.

Meanwhile Sadan had instructed both grooms to find a horse that he could buy. Greedy for the commission that they would get, the grooms looked around diligently for a horse, and soon found one. An Englishman by the name of Digby, who was an officer in the army, was returning home. His horse was up for sale. Sadan went to see it. He observed its gait, rode it, and was captivated. He came home and told Padam Singh, 'Come and see the horse. I have seen it and I like it very much. It is gentle and graceful.' Padam Singh could no longer put off the decision. He went and saw the animal, met its owner, and asked about the price. The deal was settled at four hundred rupees.

But where to find the money? There could be one or two hundred rupees lying in the house but those would be with Sobhdra, and from her he did not expect any sympathy in this matter. He had a friend, Charo Chand Chatterji, who was the manager of a bank. He thought about asking him for a loan, but never having borrowed before, he could not bring himself to do so now. He feared that his request might be turned down, and he dreaded refusal. He was altogether ignorant of this facet of life. How did people manage to overawe moneylenders? He did not know. Many times he sat down with pen and ink in order to compose his request, but couldn't think what to write.

Meanwhile, Sadan brought home the horse from Mr Digby's. The cost of the equipment was an additional fifty rupees. The payment was deferred to early next morning. For a man of Padam Singh's position and status, finding the amount was not difficult, but he could only see darkness all around him and he realized his natural weakness. A man who has never climbed heights feels giddy if he has to go up to the roof. So it was with Padam Singh. In this condition he could think of no support but Sobhdra's.

When Sobhdra saw the unhappy expression on his face, she enquired, 'You look sluggish today. Do you feel well?' Padam Singh bent his head and replied, 'Yes, I am well.'

Sobhdra: Then why are you so dejected?

Padam Singh: I don't know what to say. I am worried on account of Sadan. For days he had been insisting on buying a horse. Now he has gone and bought himself one from Mr Digby. I have to find Rs.450 to pay for it.'

Sobhdra exclaimed in surprise, 'All this happened and I wasn't told!'

'I was afraid to tell you,' Padam Singh answered guiltily. Sobhdra said with feigned sympathy, laced with sarcasm, 'There was no need for fear. I am not Sadan's enemy that I would burn up with envy. After all, now is the time for him to enjoy life. Besides it is not such a large sum. As long as you are alive you can earn this amount many times over. The boy will be pleased and that is what matters. After all he is the son of the brother who brought you up and educated you.'

Padam Singh was prepared for the sarcasm. That is why when he related the incident to Sobhdra he had woven into it complaints about Sadan's obstinacy and thoughtlessness. In reality he did not find Sadan's behaviour in the matter as unreasonable as he found his own inability to provide the money regrettable. But in order to enlist Sobhdra's sympathy he needed to win her over. He entered the den of the lioness in order to subdue her. 'I was afraid to tell you,' he said shyly, 'But I must say that though it is good to see children enjoy themselves, one must first possess the means to bear the expense. I have been agonizing all day over what I can say to Digby's man who is arriving early tomorrow morning. If only I had fallen ill—that would have given me an excuse to put off seeing him.'

Sobhdra: That's not difficult. Just stay in bed and I'll tell him you are not well.

Padam Singh could not help laughing at this. There was such indifference, such detachment and irony in what she said!

'All right, but if you send him away tomorrow, can't he return the day after? Digby is going away. We'll have to find a solution tomorrow.' He told her

Sobhdra: Why not find one today, then?

Padam Singh: If I could think of a solution, why would I have turned to you? I am at my wit's end. Tell me what I should do.

Sobhdra: How can I help you? After all, it is you who has read law. My poor brains are not equal to the task. All I know is that your enemies will be petrified when they hear the horse neighing at your door. The whole town will be buzzing with the news, and when you see Sadan mounted on the steed your eyes will be dazzled.

Padam Singh: Well tell me how all this can be achieved.

Sobhdra: Trust in God. He'll find a way for you.

Padam Singh: You are taunting me again.

Sobhdra: What else can I do? If you think I have the money you are wrong. I can't dissemble. Here are the keys of the trunk. There are a hundred or more rupees in it. Take it all, but find the rest yourself. You have so many friends, can't they help you out?

Although this was the answer Padam Singh was hoping for, he was depressed to hear it. The strength that he hoped to derive by discussing the matter with his wife had eluded him. She was not to be mollified. Harmony had not been restored. Silently he looked at the sky, like somebody drowning in a bottomless river.

Sobhdra was ready to give him the key of the trunk and if he had opened it he would have found in it not one hundred, but full five hundred rupees in a small silk purse. This amount constituted Sobhdra's savings for a full year. It used to gladden her heart just to look at this money. She would dream of buying with it a saree each for the women in her village when she visited her paternal home. Or, she thought 'If Padam Singh is ever in need of money, I'll hand him the entire sum; how happy he will be, and how surprised!'

It is unusual for beautiful women to have such lofty aims. They save money to buy jewellery. But Sobhdra came from a very affluent family. Her desire for jewellery was satiated. Nor was she tight-fisted by nature, though she had an aversion for wasteful expenditure. Now, seeing her husband helpless and downcast, she relented. She said, 'You have invited trouble unnecessarily. It was all very simple. You could have told him, 'There is no money for the horse just yet, so content yourself with the phaeton for the present.' Today he wants a horse, tomorrow he will demand a motor car. What will you do then?

Even if it were your duty to satisfy his whims, you must consider what you can afford before you commit yourself. When your brother hears of this he will not be pleased.' Saying this Sobhdra got up and taking out the silk purse from the trunk, thrust it before Padam Singh. 'There are five hundred rupees here. Take them and do what you like with them. Had they remained here they would still have been at your disposal. But anyway, take them away. At least, this money will dispel your worries. There is not a penny left in the trunk now.'

Padam Singh was stupefied. He threw a sidelong glance at the money but did not rush at it. His face showed that the burden of his worries had lightened somewhat, but the childlike excitement and pleasure that Sobhdra had hoped to see was not there. In a moment, even the shadow of satisfaction had passed out of his expression, leaving behind a look of remorse and anxiety. He was thinking, 'God knows why the poor thing saved this money; what necessities she renounced in order to save it. These are not rupees, they are her sacrifices, her murdered needs. It would be unacceptable and unjust to touch this money.'

Seeing him look disturbed, Sobhdra asked him, 'Well, doesn't this windfall make you happy?'

Padam Singh looked at her with gratitude and said, 'How can I be happy? You should not have brought out this money. I will go and return the horse. I will say it is inauspicious, or find some other fault with it. Let Sadan sulk if he likes.'

Had Sobhdra suggested returning the horse before she gave him the money, he would have been indignant. He would have considered it beneath his dignity to break his word, once given. And of course he would not have spared her for suggesting it. But her sacrifice in handing all her savings to him, had overwhelmed him. The question now was 'Should charity begin at home or outside it?' He had decided that first, he must do the decent thing at home. But when do we care for those nearest to us if it is a question of impressing the world?

Sobhdra asked him in surprise, 'How come you have changed your mind so fast? It would be wrong to return the animal now. Even if Mr Digby agrees to take it back it would be an injustice to him. He is about to leave; to return the animal at this stage would be a cheap thing to do. Take the money and give it to him. After all, money comes and goes, and the whole purpose of saving it is to use it when it is needed. I mean it when I say that I am not at all sorry to part with it.

I am giving it to you with pleasure, and if you still have reservations
then consider it a loan which you can pay back to me.'

This put the transaction in a different light. Padam Singh was very
pleased. He said, 'On that condition I will accept the money. I am
even ready to pay a reasonable interest on it, and I will return it to you
in monthly instalments.'

– 15 –

The saints and ascetics of yore prescribed two ways of cleansing
the soul. One was through friendship and amity, the other through
abstinence and restraint. These are difficult conditions, and the culture
of our cities has made them even more so by providing a host of
tantalizing temptations. It expects the individual to be like the lotus
flower which lives in the water yet keeps itself dry.

The different stages of life are dominated by different
preoccupations. Childhood is the age of sweets, and old age the time
when greed is strong. In youth, desires and ardour prevail. It is the
time of life when the marketplace* with its many attractions stirs up a
storm of temptations. Those who are steadfast, and like the lotus
flower, know how to stay dry, keep their balance. The others slip and
fall.

We try to keep the wine shops outside the precincts of the town.
We hate gambling dens too. But we house the dancing girls and
prostitutes in the decorated upper storeys of the marketplace. What is
this if not incitement? We have given the status of harmless
entertainment to the occupation of selling one's virtue. There are so
many trivial things in the market that we buy avidly, how can we
expect the human heart not to die for a precious thing like beauty? We
fail to comprehend this! A disputant would say, 'These are spurious
cravings. Thousands of young men visit the marketplace day and night,
but hardly any succumb to what you say is the ineluctable temptation.'
The disputant wants visible evidence of destruction of character, but
he does not know that though weakness of character is as prevalent as
air, it becomes known only when it is exposed. Why are we so
shameless and cowardly? What is the reason for our weakness of
spirit? Why is there such divergence between our utterances and our
actions? Why is our will so weak? Why have our values fallen so low?
All this is a manifestation of the decline of character.

Many months passed. The monsoons arrived. There was the flurry and excitement of many fairs. Sadan, dressed up like a dandy, rode around on his steed. In his heart there burned a flame of desire. The tumultuous waves of Beauty's ocean, inviting female faces behind bright glass, played havoc with the boat that was his heart. He had now become bold enough to dismount in the marketplace, and sit in the paan shops, eating the betel leaf. The shopkeepers, who thought he was the spoilt son of a rich man, regaled him with news of the prostitutes. They told him which of them sang well, which one was incomparably beautiful, who was more seductive, and which courtesan was cruel and unfaithful. Sadan would listen eagerly to this gossip. His taste was now growing more susceptible to the evocative; poetry which had seemed meaningless to him before, now sent tremors through his heartstrings, and the romantic notes of a song would mesmerize him. It was with difficulty that he kept himself from visiting the courtesans' salons.

Padam Singh wanted Sadan to be fashionable but his tendency towards rakishness disturbed him. He himself liked to enjoy the open air, but never during his daily outings did he see Sadan in a park or a meadow. However, he did see him standing in the marketplace once or twice. As soon as Sadan saw him he would immediately go into a shop, pretending to buy something. Padam Singh would see him and go away with his head bowed. He wanted to stop Sadan from coming here, but was too embarrassed to do so. The thought that Sadan had been affected by the licentious atmosphere of the marketplace began to trouble him.

One evening, Padam Singh had gone out to take the air. Two gentlemen, one, by the name of Abul Wafa, and the other, a certain Abdul Lateef, who were out in their phaeton, stopped and greeted him.

Said Abul Wafa, 'Sir we were just talking about you. Why don't you ride with us for some distance.'

Padam Singh: Please excuse me. I am used to going out for a walk in the fresh air at this hour.

Abul Wafa: There is something we want to tell you, and we were about to come to your house for this purpose.

Abdul Lateef: You will greatly appreciate the information.

Padam Singh was obliged to climb into their phaeton.

Abul Wafa: If you promise us a reward we will give you a wonderful and thrilling piece of news.

Padam Singh: Well, tell me.

Abul Wafa: Your ex-cook has transformed herself into Maharajan Saman Bai.

Abdul Latif: By God, we admire your taste! She made her appearance in the courtesans' district only three or four days ago, but in this short time she has, like the full moon, eclipsed all the stars. When she is present, none of the others have any appeal. Across from her abode there are throngs of profligates. Her face is a rose and her body is of pure, lustrous gold. Sir, on my word, I've never before seen such a captivating face.

Abul Wafa: Her gestures are enchanting too. A man who can claim to keep his virtue after seeing her, will have my undying allegiance. Only somebody as discerning as you could have plucked a gem like her from a heap of rags.

Abdul Latif: She appears to be extremely intelligent too. It couldn't be more than five or six months since she left your house, but yesterday when we heard her sing we were astounded. There is no one in this town whose voice has so much purity and delicacy.

Abul Wafa: She is the talk of the town. People are spellbound by her. Balbahadar Das has been seen to frequent her. Let's go there. You can claim admittance on the grounds of your long acquaintanceship, and we will benefit on account of your old ties.

Abdul Latif: We are going to take you there now. You can see her in private whenever you like, but for the present you must oblige us.

Hearing their talk, Padam Singh was so overcome by regret and shame that he could not lift his head. His apprehensions had proved right. He wanted to ponder over this tragedy in solitude, to establish how far he himself was responsible for it. To his two importunate companions he said, 'Please excuse me. I will not be able to go with you.'

Abul Wafa: Why?

Padam Singh: Because I don't like to see a woman from a decent background in such circumstances. You can think what you like but my only connection with her was that she used to visit my wife.

Abdul Latif: Sir, put off the sermon. We have spent a lifetime in that environment and are familiar with every bend in the road. Just introduce us. Your intervention will be useful to us.

Padam Singh said impatiently, 'I have told you that I will not go there. Let me get off.'

Abul Wafa: And we have said that we will take you there. You will have to take the trouble for our sake.

Abdul Latif spurred on the horse with a lash from his whip. Padam Singh said angrily, 'You want to humiliate me?'

Abul Wafa: Sir, one should accommodate one's friends. We'll soon be there. Here we are at the turning.

Padam Singh understood that the two men were bent on making mischief. They would not heed him. Rather than go to Saman he would have preferred to jump into a well. Deciding quickly what he was going to do, he got up and jumped out of the fast-moving vehicle. He tried to steady himself but could not, and sprinted giddily for about fifty feet before he stumbled and fell. His elbows were badly hurt and he was drenched in perspiration. Faint with vertigo and gasping for breath, he managed to sit up on the ground. Abdul Latif stopped the horse and both men came running to him. They fanned him with a handkerchief and in fifteen minutes he felt better. The two men expressed their remorse and begged to be pardoned. They insisted that they would take him to his house in their phaeton but on no account would he agree to this. Finally they left, and he began to limp home. When he was calmer he was surprised at himself for jumping out of the moving vehicle. 'Had I told them sternly to stop the phaeton they wouldn't have dared to disobey. And if they had still not done so I could always have snatched the reins from their hands. Anyway, whatever happened was for the best. If they had diverted me with their talk and I found myself at Saman's door, that would have been difficult. I could not have faced her. Perhaps when I dismounted from the phaeton there, I would have run through the marketplace like a madman. I can see the throat of a cow being cut, but I can't see Saman in her present circumstances.'

Our greatest fears are the imaginary ones.

The question he kept asking himself was, who was responsible for Saman's transformation? His judgement was based on his analysis of what had taken place in the past. 'If I had not turned her away from my door she would not have been ruined. After she left my house she could find no sanctuary, and out of anger and unhappiness she took to prostitution. I am to blame for this.'

'But why was Gujadhar so angry with Saman? She was not a woman in purdah, she went to public places like fairs. Just for coming home late, on one day, he should not have punished her so cruelly. He should have rebuked her, he could even have struck her. Saman would have cried and his anger would have subsided. He could then have appeased her, and that would have been the end of the matter. But this did not happen, because Bithal Das had fed his wrath. To be sure it is Bithal Das who is to blame. It was he who spread rumours about me so that I was obliged to be harsh and turn her away.' Reinterpretting past events in this way, Padam Singh succeeded in pinning the whole blame on Bithal Das, and found this comforting. This conclusion also helped to fan the spark of malice and retaliation that he had been harbouring in his heart for some time. He had found a way to humiliate Bithal Das. As soon as he got home, he sat down to write a note to Bithal Das, forgetting even to change into his indoor garments.

Dear Sir,

You will be happy to hear that Saman has now made her appearance in the courtesans' quarter. You will perhaps remember that on Holi she had taken refuge in my house from the fury of her husband, and out of compassion, I had thought it appropriate to let her stay for a few days, until the wrath of her husband had subsided. But during this short period, some of my friends who were not at all ignorant of my character, started to disgrace me. The upshot was that I was forced to turn that unfortunate woman out of my house. And finally she met the fate that I had feared she would. I hope that a review of these events will make it clear to you whether my intentions were guided by humanitarian or villainous considerations.

Regards,
Padam Singh.

– 16 –

Baboo Bithal Das was the soul of all the local movements and other activities in the city. No project was completed without his assistance. Others started movements which then became his responsibility to execute. This courageous man shouldered such burdens cheerfully. He never complained of the pressure of work. He barely had time to eat, and none to relax at home. His wife accused him of negligence, his sons roamed about, free of all discipline. But Bithal Das had sacrificed all personal interests at the altar of public good. He was either collecting donations for orphanages, or finding funds for the education of poor students. During epidemics of cholera and the plague, his self-sacrifice became superhuman. During famines he would go from village to village carrying grain on his head. Recently, when the Ganges was in flood, he did not go home for months because he was busy collecting donations, making inventories of the damage, and distributing aid. He had given away many of his possessions to the needy, but never boasted of it. He had not received higher education, and he was not very eloquent. However his views had maturity and foresight. And although he was not an organized, clever, or alert person, the quality he had of caring for humanity, made him a respected and outstanding citizen.

When Bithal Das read Padam Singh's letter it was as if he had been slapped. He was in the habit of looking at the practical side of every situation. He was not vexed at the tone of the letter, nor did it trouble him to learn how, through his misjudgement, he had hurt an innocent friend. He was not in the habit of wasting time on regrets. His concern was, what action was needed in the present. Once he had decided that, he was not given to vacillation. So, Bithal Das put on his clothes and went straight to the marketplace. He found Saman's house, went upstairs, and knocked at her door. Hira, who was the mistress of Saman's brothel, opened the door.

It was past nine o'clock and Saman was about to go to bed. She was startled when she saw Bithal Das. She had seen him many times in Padam Singh's house. She had heard about him from Lala Chaman Lal and Abul Wafa. They had described him as impudent, an enemy of dance and music, a dry ascetic, and a destroyer of brothels. She therefore mistrusted him. She stood there uncomfortably and said, 'What brings you here, sir?'

Bithal Das sat down informally on the floor and said, 'I have come here for a purpose. I see here what I had found impossible to believe. When I received Padam Singh's letter I thought that the account he had received was exaggerated. But how can my own eyes deceive me! Saman, you have made the whole Hindu nation hang its head in shame!'

Saman replied soberly, 'Perhaps you think so, but nobody else does. A little while ago there were many people here listening to the singing. None of them were hanging their heads in shame. In fact they all seemed very happy at my arrival. Again, I am not the only Brahmin in this business. I can name a few who come from important families. I couldn't get along at home so was obliged to come here. If the

Hindu nation itself has no shame how can helpless women like me help them?'

Bithal Das: Saman, you are right. It is true that the Hindus have fallen very low. They would have disappeared without leaving a trace, but it is the pious Hindus who have kept them alive. It is their love of virtue and self respect that brightens the face of the Hindu nation. Just to preserve their honour, hundreds of thousands of their women jumped into the fire.* This is the sacred land where women bore unspeakable cruelties and put up with humility, yet by never uttering a word about their men's brutalities they protected the honour of the Hindu nation. And these were the qualities of the common Hindu women. Those of the Brahmin woman are indescribably great. But how shameful it is that the same women are now blackening the name of the Hindus. Saman, I agree that you were very uncomfortable in your home, that your husband was poor and bad-tempered, and that he callously threw you out of your home, but a Brahmin woman tolerates all this for the sake of her family and her caste. To bear misfortune with fortitude is the obligation of a Brahmin woman. But you did what the shameless women of low castes do. When they are angry with their husbands they go back to their own families and when they find that they can't cope there, they turn to prostitution. Just think how shameful it is that the circumstances in which a multitude of your sisters live and put a brave face on, you have found so unpleasant, that you destroyed your own honour and the dignity of your family to take this route.

'Have you not seen women who are much poorer and more unfortunate and helpless than yourself, but in whose heads the thought of taking such a step never entered? If it had not been so, this sacred land would have been worse than hell. Saman, this act of yours has humiliated not just the Brahmin caste but the whole of the Hindu community.'

Saman's eyes filled with tears and she could not lift her head for shame. Bithal Das continued, 'No doubt you live here in luxury. Your abode is a well-adorned palace, you sit on soft carpets, sleep on a bed of roses, eat exotic delicacies. But only think, you have bought this luxury by selling your honour! How well-respected you were, people looked at you with reverence; but now it is sinful to look at you!'

Saman interrupted this speech to say, 'Sir, what are you saying! My experience has been that there was no question, then, of the respect that I now get. Once I went to Seth Chaman Lal's Thakar dawara to see the swing festival.* I stood in the rain all night but nobody let me in. But yesterday when I performed in the same place it seemed as if my feet had blessed that temple.'

Bithal Das sat up and said, 'But did it occur to you what kind of character such people have?'

Saman: I don't know that. But I do know that he is the chief of the Benaras Hindus. But it is not just he, all day I see thousands of people on this street, educated, uneducated, rich, poor, wise, ignorant—I can see them all. All of them stare at me, overtly or covertly. There isn't one among them who will not go out of his mind with happiness if I throw him a look of partiality. There may be two or three people in the whole city who despise me. One of them is you, and your friend Pandit Padam Singh is another. But since the world respects me why should I care for the disapproval of a handful of people? Even Padam Singh hates me, personally; he does not hate the rest of my community. With these very eyes I have seen him divert himself with Bholi.

Bithal Das knew how to attack, but did not know how to defend his arguments. He could not think of an appropriate reply and so was rendered speechless. Saman continued, 'You must be thinking that I entered this business so that I could enjoy a life of luxury, but you are utterly mistaken. I am not so blind that I can't tell the difference between good and evil. I am conscious that I am guilty of a shameful act. But I was helpless. For me there was no other way. I can tell you my story if you care to hear it. I am sure you realize that everybody in the world does not have the same temperament. Some people can put up with indignity better than others. I belonged to an upper class family, but due to the folly of my parents I was married off to a poor and uncouth man. However, even in poverty I could not tolerate insults. When I used to see those people elevated, who deserved to be despised, I used to seethe, though only inwardly. I never complained of my bad luck to anybody. It is likely that this anguish would have gradually burnt itself out. But the party at Padam Singh's on Holi refuelled the flames of the anger within me. You know the rest of the wretched episode. After I left Padam Singh's house I was trapped by Bholi. But even at that stage I tried to avoid this evil route, and tried to earn a living by sewing garments. However the rakes would not let me be, so finally I had to take the plunge and that is how I arrived at this house of infamy. It is extremely difficult to remain pure in this place, yet I have vowed to remain so. I will sing, I will dance, but I will stay chaste, and God willing I will be able to keep my vow.'

Bithal Das: The very fact that you are here is enough to defame you.

Saman: What else can I do? Tell me of some other way by which I can spend my life comfortably.

Bithal Das: If you think that you will always be comfortable here, you are wrong. You will realize in two or three years that comfort comes with contentment. Luxury alone never confers comfort.

Saman: Well, if I can't get comfort, at least I am respected here; I am not a slave.

Bithal Das: There again you are wrong. You may not be the slave of another person, but you are the slave of your own desires. Being enslaved by one's desires is much worse than the other kind of slavery. Here you will get neither comfort nor respect. Yes, you will enjoy the luxury for some time, but finally you will lose that too. Just think, what an injustice you are doing to yourself and to your community for the sake of living in luxury for a few days.

Saman had never heard such ideas from anyone. She had always thought that self interest was the purpose of life. The well-being of self and outward dignity were the two principles which governed her life. Only now had she become aware that tranquillity and real dignity were goods that could only be found in the shop of contentment. She said, 'All right, I will renounce both these things. But what should I do for a living?'

Bithal Das: You need not turn to this place for a living. There is so much you can do sitting at home.

Saman could not find another excuse. Bithal Das' faith had vanquished her. We cannot deceive an honest man, because his honesty inspires us with noble sentiments. She said, 'I myself am ashamed to live here. Tell me what alternative I have. I am good at singing. Perhaps I can give singing lessons.'

Bithal Das: There is no singing school for girls here.

Saman: I am educated and can teach well. I could teach girls.

Bithal Das answered in a worried tone, 'There are many girls' schools but it is unlikely that they will offer you a place.'

Saman: Then, what else can you suggest? Is there a friend of the Hindu nation who can give me fifty rupees a month to live on?

Bithal Das: That's difficult.

Saman: Then do you want me to grind grain for a living? I am not that humble!

Bithal Das said with embarrassment, 'If you want to live in the Widows' Ashram, it can be arranged.'

Saman replied thoughtfully, 'Yes, that is acceptable. But if the women there taunt me I will not stay for a moment. I cannot suffer such humiliation.'

Bithal Das: That is a difficult condition. How can I stop people from talking? But I think that the management there will not agree to take you in.

Saman said sarcastically, 'If your Hindu nation is so insensitive why should I make sacrifices for its sake? Why should I give my life? If you cannot persuade your nation to own me and if that nation has no shame left, then how is it my fault? I will make just one more suggestion and if you reject even that, then I will not bother you anymore. Please bring Padam Singh here just for an hour. I want to tell him something in private. If you succeed I will leave this place immediately. I just want to know what value the person who you think loves his people, places upon my return to modesty.'

Bithal Das replied happily: Yes, this I can do. Tell me when you want him to come?

Saman: Whenever you like.

Bithal Das: But you will keep your word?

Saman: I haven't fallen so low yet.

– 17 –

Public servants are rarely rewarded with perfect victories. They are satisfied with partial successes. Bithal Das was overjoyed, as though he had found buried treasure. He was certain that Padam Singh would think nothing of this slight inconvenience and would agree to meet Saman as soon as he was told of the condition. He had not been to see Padam Singh since before Holi. He had spread false rumours about him and perhaps he regretted this. However his embarrassment did not stop him now. It was ten o'clock at night when he started for Padam Singh's house. The sky was clouded and its dark shadow fell on the ground. But in the courtesans' quarters there was much activity. In their parlours there was much whispering. Sweet music as well as carefree laughter could be heard. Naked intoxication, unveiled, seemed to be out on a jaunt in the district.

Once he left the marketplace behind, it seemed as though he had left behind a garden and arrived in a desolate wilderness. He met two acquaintances whom he at once grabbed and gave the news of his victory. He said, 'Do you know where I am coming from? I went to Saman Bai's place, and the spell I cast on her has put her in my power, completely. She is now so ashamed of herself that she has agreed to live in the ashram. This is the way to do a job! She tried her best and argued all the way, but finally she was overwhelmed.'

Padam Singh was in his bed, but he was not yet asleep when Bithal Das called.

Jitan was sitting in his room counting the day's earnings when he heard Bithal Das's voice. He quickly hid the money somewhere around his waist and called out, 'Who's there?'

Bithal Das: It's me. Is Padam Singh asleep? Go and wake him up. Tell him I am standing outside. I've come on urgent business and can he step out.

Jitan was very irritated. His calculations were incomplete. God knows how much more money there was left to be accounted. Reluctantly he got up and informed the Pandit of the caller. Padam Singh realized that something new must have come up for Bithal Das to visit him so late at night. He came out at once.

Bithal Das said, 'Forgive me for troubling you. Do you know where I am coming from? I was at Saman Bai's. I went to see her as soon as I got your note. It was not only her reputation that is at stake, it is the reputation of the whole Hindu nation. Anyway, I got there and was astounded to see the luxury that she lives in. That simple girl is now the queen of the red-light area. When did she learn such finesse? Perfect enunciation, fascinating style. I harangued her. At first she listened silently to me, then she began to weep. I knew at once that the iron was hot, and all I needed to do was to strike once or twice. And that really succeeded. She was outwitted. At first she was alarmed at the mention of the Widows' Ashram, she asked to be given fifty rupees a month to live on. But as you know there is nobody who will part with a sum like that, so I didn't agree. Finally, after much persuasion she agreed to go to the Widows' Ashram, but on one condition. Now it is up to you to fulfil that condition.

Padam Singh looked at Bithal Das wildly.

Bithal Das: Don't worry, it is a very simple condition—just that you should go and see her. There is something she wants to tell you. I knew you would have no objection in complying, so I agreed to bring you. Now tell me when you can come? I think we should go early tomorrow morning.

Padam Singh was as slow to make a decision as Bithal Das was quick. It took him hours of weighing pros and cons before he could make up his mind about a certain course of action. He began to think, 'What is the meaning of this condition? What does she want to say to me? Couldn't it be said in a letter? There is something suspicious about this. Abul Wafa must have told her what happened today, and as she is in an arrogant mood these days she must have thought "He will have to come here and I will see that he does!" So, she just wants to humiliate me. And what if she doesn't keep her word about going to the ashram, even if I go?' This last thought seemed like a good excuse to Padam Singh and he voiced it to Bithal Das:

'What if she doesn't keep her word?'

Bithal Das: Is that likely? Do you want a written contract?

Padam Singh: It's not a question of a contract, I feel that she would not be inclined to leave her luxurious life and go to live in the ashram. In any case the management is bound to refuse.

Bithal Das: It is my job to convince the management. And if they don't agree, I'll find some other means of livelihood for her. But in the first place, even if she doesn't keep her word we have nothing to lose. By your going there at least we would have done our duty.

Padam Singh: Well, the satisfaction of doing one's duty can be had; but you will see, she is bound to deceive us.

Bithal Das became impatient. Padam Singh needed to be treated with fortitude at that moment, but instead, he said to him sharply, 'Even if it is a deception, what do you lose by going there?'

Padam Singh: Perhaps you don't care for my reputation, but I don't consider myself so abject.

Bithal Das: In short, you will not go?

Padam Singh: My going there will serve no purpose, but if the idea is to humiliate me, that's another matter.

Bithal Das: What a pity it is that you are so unwilling to do anything for such a noble cause. Alas, you can see that a Hindu woman has fallen into a well, and yet you who are an enlightened member of the same faith are reluctant to rescue her! All you are fit for is to suck the blood of poor ignorant farmers,* and nothing more!

Padam Singh made no answer to this reprimand. In his heart he acknowledged that he had no courage and believed that he deserved the rebuke, however he was displeased at it being voiced by a person who he thought was one of the main perpetrators of Saman's tragedy. It caused him an effort not to make an angry retort, but the truth was that he wanted to save Saman, though in a less overt fashion. He said, 'There must be other conditions?'

Bithal Das: Yes, but can you satisfy them? She wants fifty rupees a month to live on. Can you give that?

Padam Singh: Not fifty, but I can give twenty.

Bithal Das: O come on! You can't stand a bit of inconvenience and you think you will spare twenty rupees a month.

Padam Singh: I promise you that I will pay twenty rupees a month, now. And if there is an increase in my income I will not hesitate to pay the full sum. At present I am rather constrained for money. Even the twenty rupees can be procured only if I sell the horse and the carriage. I don't know why but my business is slow these days.

Bithal Das: Even if you found the twenty rupees, where will the rest of the money come from? You know what the others are like; even the subscription for the ashram is not easy to collect. Anyway, I'll go now and I'll try my best, but if I don't succeed it will be all your fault.

– 18 –

It is evening. Sadan, mounted on his horse, is slowly passing by, gazing at two parlour windows of the red-light district. Ever since Saman had arrived here it became Sadan's daily practice to stop in front of her quarters, on some pretext or the other. The beauty of this young blossom had enticed him to the point that he found no peace at all. There was a simplicity and reserve in her beauty that affected him more than pleasing gestures and amorous glances. He would have liked to devote his love to this personification of beauty, but he could not get a chance to do so. Saman was always surrounded by throngs of people who were the worshippers of beauty. Sadan was afraid that there would be friends of his uncle's among them. This thought discouraged him from climbing the stairs to where she lived. Hiding a flood of desire in his breast, he would go away disappointed every day. But on that day he resolved to meet her, regardless of how long he might have to wait. He could not bear the frustration of staying deprived any more. He arrived in front of Saman's dwelling. The lovely sound of a classical air was coming from its direction. He rode on, and after killing time for two hours in the park and the marketplace he returned to the district at nine. The silver moonlight had covered the high roofs and parapets of the area with a sheet of radiance, and the market where beauty was traded had become the embodiment of beauty. Sadan arrived again in front of Saman's apartment. The music had stopped and no voices could be heard. When he became certain that the coast was clear, Sadan dismounted, and tied his horse to a pillar of the shop below. He was breathing hard and his heart was beating against his chest.

Saman's performance had just come to an end and she was feeling in her heart the sadness that is like the silence which follows a storm, and which is the last stage of a session of pleasure and gaiety. This sadness is a message from the Unseen which momentarily awakens those who are intoxicated with pleasure. The past appears before them like a pleasant scene in a dream. For a while the eye of our unconscious mind opens, and the light of thought illuminates a dark corner of it. Saman was thinking of Sobhdra. She was comparing herself to her. 'Will I ever know a peaceful life like hers? Impossible! This is a market where the goods are lust and display. No peace can be found here! When the time for Padam Singh to come home from the courts drew near, how happily Sobhdra would prepare his paan. She would cook fresh halwa for him. And when he arrived, with what ardour she would greet him! I even saw them embrace. What pure love that was, and how full of happiness! And I? Only the blind come here, or the boastful. Someone spreads the web of his wealth, another of his hypocritical chatter, and yet another of his imaginary love! They are dry, unfeeling, lifeless leaves, all of them, colourful and pretty but not affected by spring or autumn, summer or winter. They may appeal to foolish children, but no sweetness-seeking insects will ever hover over them, no nightingale will sing around those coloured leaves.'

Sadan entered the room. Saman started. She had been watching Sadan come and go for many days. His face resembled Padam Singh's, though instead of pale sobriety it reflected animated dandyism. There was no trace of meanness, haughtiness, or shallowness, which were the distinguishing qualities of those who came to pluck the flowers of this garden. He looked like a simple, and sincere young man. Saman had seen him that day gazing intently at the salons. She could tell immediately that the pigeon was getting ready to fly and would soon alight on a roof. When she saw him in front of her she experienced the proud delight which the victorious wrestler feels when he has downed his rival. She got up with a smile and extended her hand to Sadan.

Sadan's innocent face went red. His eyes lowered and he seemed overwhelmed. His tongue wouldn't utter a word. It was as though a man, adept at diving, had begun to flounder on finding himself in the water.

– 19 –

S adan had tried to hide his identity from Saman Bai. He had given his name as Kanwar* Sadan Singh. But he could not keep his secret for very long. Saman had discovered the truth with the help of Hira. This put her in a strange predicament. On the one hand she could not rest until she saw Sadan. Her heart was drawn to him more every day. No raja or other rich man found it easy to be admitted into her presence as long as Sadan was with her. She had dropped the coy reserve of the sought after, and from being the beloved had turned into the lover. But the romance did not progress beyond looks and speech because she considered this love illicit and kept it a secret even from Hira. On the basis of her old friendship with Padam Singh and Sobhdra she had formed an imaginary delicate relationship with Sadan and although this relationship had no existence outside her imagination, it acted as a chain that bound her longing. 'What if Padam Singh and Sobhdra found out! What would they think of me, and how unhappy it would make them! They would hate me for it.'

If sometimes Sadan became too bold in his speech she would change the subject. If Sadan's fingers became inclined to mischief, she would look at him lovingly and gently push his hand away. But nevertheless she wanted to keep Sadan involved, because besides her own yearning for him she was afraid that if he stopped seeing her he might begin to frequent another, and then it would not be easy to disentangle him. She saw him as a kind of trust, that had to be protected from others. Her relationship with him was therefore no more than a cautious and limited transgression.

Sadan considered this abstinence to be her due. His innocent heart had been invaded by a deep love. Saman had become the most important part of his life, but strangely enough, he managed to curb his desires in spite of his longing. His rough obstinacy had disappeared and he wanted to do whatever she wished. Had Saman forbidden him to see her he would have committed suicide, but he would not have refused to comply. The selfishness that comes to the fore in commercial love had, under the influence of genuine feeling, bound itself to the goal of seeking approval. But seeing that her abstention only grew stricter by the day he began to feel that his love was not valued.

However he had so much confidence in his manly attraction that he continued his visits. Soon he began to feel the need for bringing gifts, but how to find the means? On what pretext to ask for money? After pondering over this problem for a long time he wrote a letter to his father. In it he said, 'The food provided here is not to my liking. I feel embarrassed to speak to Uncle about this, so please send me some money.'

When the letter reached his family, Bhama began to taunt her husband, 'You were so proud of your brother. What do you think of him now? As for Sadan, he thought the world of his uncle; his eyes must have opened now. People don't remember what you do for them, anymore. I never used to spare myself for him, and this is what I get for it. But it is not the fault of my poor boy. It is all that lady's doing. When I see her I'll give her a piece of my mind!'

Madan Singh suspected that Sadan had found this ruse in order to get some money—he had full confidence in his brother. However on Bhama's insistence he had to send the money.

Meanwhile Sadan was going to the post office every day to check whether the money had arrived. Finally, on the fourth day he received twenty-five rupees. Since the postman knew him, he handed the money to him without demur. Sadan was overjoyed. That very evening he went to the market and bought a fine silk saree. However, he remembered that since he had appended 'Kanwar' to his name for her benefit, Saman must think him a prince. She might find the saree too cheap, not good enough. With the saree in his pocket he was riding back and forth on his horse. He was feeling too embarrassed to go up to her with this present. Finally, when it became very dark he braced himself and went up. He took the saree from his pocket and put it furtively on the dressing-table.

Saman had been anxiously waiting for him and was delighted when he arrived. 'What have you brought?' she asked him.

Shyly, Sadan replied, 'I saw this saree. I thought it was pretty so I bought it.'

Saman: You made me wait so long. Is this a peace offering?

Saying this she examined the saree. It was an expensive one. Questions arose in her mind, 'Where did he get so much money? Did he pilfer it from the house? Why would Padam Singh hand him such a large sum? He must have asked for the money on some pretext, or just taken it.' At first Saman thought she should return the saree but decided against it for fear of offending Sadan. At the same time she was afraid he might bring more presents, so she decided to accept the saree but warn him against repeating the gesture. She said to him, 'I am grateful for your kindness, but I am not hungry for presents. You do me a favour by taking the trouble to come here. All I want from you is a look of love.'

But when even after giving the present Sadan found that Saman continued to deny herself to him he was disconcerted. He thought, 'I gave her a humble gift and am expecting so much in return! While others shower her with pearls and precious gems and yet do not achieve their goal, I expect to succeed with just my abracadabra!' He began to think of some way of buying her a more valuable gift, though for months after, he found no opportunity to achieve his object.

One day, when about to take a bath, he found that there was no soap. He went to the other bathroom to look for it. As soon as he entered, his eyes fell on the shelf. A bracelet was lying there. Sobhdra had just taken a bath there. She had put her bracelet on the shelf before her bath. Since the time for Padam Singh to go to the courts was drawing near, she had hurried out and busied herself with cooking the meal. The bracelet was forgotten on the shelf. Seeing it lying there Sadan quickly picked it up. At that moment his intentions were not dishonest. He had thought, 'When Aunt gets really upset, I'll ask for a bribe and return it to her. It will be fun.' So he hid the bracelet in his trunk.

Meanwhile Sobhdra lay down after her midday meal. The hot weather made her sleepy and when she woke up it was evening. Padam Singh had arrived home from work and Sobhdra started talking to him. Sadan came in several times to see if there was any talk of the missing bracelet but he heard no mention of it. In the evening when he was leaving for his usual outing, it suddenly occurred to him that he could give the bracelet to Saman. 'Nobody is going to ask me' he thought, 'If they do, I'll tell them I know nothing about it. Aunt will think one of the servants has picked it up. Maybe it will not even occur to them to ask me.' And he put the bracelet in his pocket. Sometimes opportunity can change even recreation into wrongdoing.

That day Sadan found no pleasure in his outing. He was impatient to present his gift. He went to the marketplace and bought a small velvet box. He put the bracelet in it and arrived at Saman's house. He wanted to present it casually, as though for him it was an ordinary gift, of not much value. It was necessary to express humility along with the show of wealth. He sat for a long time. Saman had kept the evening free for him but he was not even inclined to talk of love. He could not think how to present the gift. Finally, when it grew late, he got up slowly, took the box out of his pocket and putting it on the bed moved towards the door. Saman noticed the box and asked him. 'What's in this?'

Sadan: Nothing. It is empty.

Saman: No, wait. Let me see.

Saying this Saman caught his hand and opened the box. She had seen the bracelet on Sobhdra's wrist, so she recognized it at once. A burden seemed to fall on her heart. Sadly, she said to him, 'I told you that I am not hungry for presents. You are embarrassing me unnecessarily.'

Sadan replied carelessly, 'A raja should accept the worthless gifts of the poor.'

Saman: My greatest gift is a look from you. You can give this one from me to your new rani. The love I have for you is innocent of such expectations. You still seem to think I am a street woman. You are the one person whom I want to love with a true, unselfish love. But even you have no use for it.

Sadan's eyes filled with tears. He thought, 'No doubt I am guilty. I have tried to buy a precious thing like her love with these worthless gifts. I have been too impatient. People in this town are ready to sacrifice everything for one look from her. The wealthiest of them come to her but she doesn't bother to even look at them. I am so mean and ignorant that I still doubt her sincerity.' This unhappy thought made him weep. Saman thought he was offended. She asked him lovingly, 'Are you offended?'

Sadan dried his eyes and said, 'Yes, I am.'

Saman: Why, what did I do?

Sadan: You hurt me. You think I want to buy your love with this rubbish.

Saman: Then why do you bring these things?

Sadan: Because I feel like it.

Saman: Spare me such favours.

Sadan: We'll see.

Saman: I'll keep this bracelet for your sake. But I keep it as a trust. You are not yet free to spend so much. When you come into your estate I'll demand what I like; I'll not let you off, then. But if you spend so much money now your family will become suspicious, and if they stop you from coming here, how will I see you?

– 20 –

Baboo Bithal Das was very thorough. When Padam Singh refused to oblige, he began to worry about procuring a monthly income of fifty rupees for Saman. Many of his projects were funded by donations, but to collect the money was always difficult. The construction of the Widows' Ashram had begun. But for two years there were no funds to put a roof over the walls. Books for the free library were being eaten up by white ants, but there was no money to build cupboards for them. In spite of these obstacles he could think of no solution but to arrange for a contribution for Saman.

Seth Balbahadar Das was one of the richest men in the town. He was also an honorary magistrate and the President of the Municipal Board. The Seth was lying on an armchair in his house, smoking a hookah. He was lean and fair, and also very well-dressed and cultivated. He had the expression of a thoughtful and intelligent man. He undertook a venture only after much deliberation. On hearing Bithal Das's proposal, he said thoughtfully, 'It is a good plan but, tell me, where will she live?'

Bithal Das: In the Widows' Ashram.

Balbahadar Das: The ashram will get a bad name, and I wouldn't be surprised if the widows leave.

Bithal Das: Then I'll find a house for her.

Balbahadar Das: There will be battles among the young men of the neighbourhood.

Bithal Das: Then what solution can you offer?

Balbahadar Das: In my opinion you should not involve yourself in this. Once inhibitions are cast off, they do not return. An afflicted limb is cut off so that it does not infect the rest of the body. The same should be true of society. I can see that you do not agree with me but I have told you frankly what I think. Moreover, I am a member of the ashram's management and I would never advise keeping an immoral woman there.

Bithal Das said with much annoyance, 'In short you will not help me in this good work. If sensible people like you have no sympathy what can be expected of the others? Forgive me for wasting your time.' Saying this Bithal Das got up and went to the house of Seth Chaman Lal.

The Seth was a dark, shapeless man. He was very tall and slack. Neither cleanliness nor order were part of his nature. Like his body his thoughts were also shapeless. They were narrow where they should have been broad and were broad where they ought to have been narrow. He was the President of Dharam Sabha,* the Chairman of the Ram Lila* Committee and its patron. He thought politics was a poisonous snake and newspapers were the fangs of that snake. He kept contacts with officials. British officials especially appreciated him. His talents were much valued in their circles. He was neither generous nor mean. Inspection of subscription lists was his guiding light. He had a special quality which hid all his shortcomings; it was his sense of humour. On hearing Bithal Das's proposal he said, 'Babu Sahib, you are a colourless person. You have no taste. It has been a long time since we had someone worthwhile in the red-light district, and you are bent on making her vanish. At least wait till after the feast of Rama's birthday. She will sing on the occasion. After all, women of the lower castes come to the temples and desecrate them, why shouldn't a Brahmin woman be allowed in? Anyway, all that was said in jest, so forgive me. It is wonderful how you are always so immersed in good works. Where is the list of donations?'

Scratching his head, Bithal Das replied, 'So far I've only approached Balbahadar Das, but you know how artful he is. He evaded the issue.'

Had Bithal Das contributed one, Chaman Lal would have had no hesitation in putting down two, and if the former had promised two, the latter would have guaranteed four. But if the figure is zero then the product can only be zero. Chaman Lal began to think of an excuse. He found one immediately, and said, 'I sympathize with you, but Balbahadar Das must have had some good reason for not agreeing to subscribe. Now that I think about it, I can see the hand of politics in this affair. Perhaps you don't look at it that way but to me clearly politics is involved. The Muslims will take offence and they will complain to the authorities. As you know, the authorities have no eyes, they have only ears. They will immediately start suspecting a conspiracy.'

Bithal Das said impatiently, 'Why don't you say frankly that you don't want to contribute. I don't think Muslims are so biased that they will mistrust this good deed. I am certain that in fact they will assist us. There is no need to give a political colour to a simple matter like this. You can give an outright refusal if you don't want to be involved.'

Seth Chaman Lal was embarrassed. He wanted to say something but Bithal Das did not give him a chance to do so, and immediately got up to go. It was not the first time he had been disappointed. He experienced such rejections frequently. Perhaps he would have been less ineffective if he had been more patient. But by nature he was impatient.

Bithal Das next went to see Dr Shyama Charan. This gentleman was highly educated and intelligent. One of the kingpins of society, he was at the height of his legal practice. A great believer in moderation, he weighed every word that he uttered. His taciturnity was considered maturity of views. He was devoted to the principle of 'Silence is Golden.'

Since he was a follower of the 'middle path' school of thought, his opposition did not harm, nor his support benefit anybody much. Everybody thought him a friend and everybody thought him an enemy. In this play of light and shadow, his reputation remained evergreen. He was a member of the local council and used to spend some of his time every day with opinion makers. When he heard Bithal Das's suggestion he said, 'My sympathies are with you in this matter and I'll do whatever I can to help you. But it is important to highlight the circumstances which give rise to such mischief. If you save this one woman what good will that do? Tragedies such as these happen here all the time. It is essential to remove the causes. If you like I can raise some questions at the council?'

Bithal Das jumped up. Raising a question at the council was for him a very important event. 'Yes, that would be very appropriate,' he said.

Doctor Sahib immediately produced a series of questions for the council:

1. Does the government know the figures for the increase in the number of prostitutes last year?
2. Has the government tried to investigate the reasons for this increase?
3. To what extent are the reasons, economic, emotional, or, cultural?

After writing down the questions, Doctor Sahib turned to his clients. Bithal Das waited for half an hour and then impatiently asked what he should do.

Doctor Sahib: Don't worry, I will definitely ask these questions at the next sitting of the council and I'll let you know the result, or you can read it in the papers.

Bithal Das wanted to take the Doctor to task but thought better of it.

This stubborn man who could find no peace until he had fulfilled his mission, then went from pillar to post, and approached different people every day. His efforts were not all wasted, as he did manage to raise several hundred rupees and was promised several hundred more. Yet he could not make up, fully, the difference of thirty rupees a month. After three months of going to and fro he had with difficulty, collected just enough for an assured income of ten rupees a month.

Finally, when he had given up hope of finding more assistance, he went early one morning to see Saman.

When she saw him she said with concealed sarcasm, 'Well sir, what brings you here?'

Bithal Das: Do you remember your promise?

Saman: After such a long time, if I had forgotten it I would be quite justified.

Bithal Das: I tried hard to find a way, but I had to deal with this unfortunate nation which has no feeling of nationhood. However my efforts were not entirely wasted, and I have managed to make arrangements for a monthly income of thirty rupees. The rest too I hope will be found sooner or later. I request you to accept this and leave for ever this ugly business.

Saman: You couldn't bring Padam Singh.

Bithal Das: He could not be persuaded to come, under any condition. But twenty rupees out of the thirty have been gifted by him.

Saman said in astonishment, 'I never thought he would be so generous. Did you get anything from the Seths?'

Bithal Das: Don't mention the Seths! Chaman Lal can part happily with a thousand or two for the Ramlila celebrations. Balbahadar Das can be even more generous in his hospitality to the authorities. But for this cause they could not spare anything.

Saman was in love with Sadan those days. She had never before known the happiness of love. And having found it she did not want to give it up. Although she knew that the relationship could only end in separation, yet she told herself, 'Why shouldn't I cling to this happiness as long as I can? I don't know what the future will bring. Who can tell when the boat of my life will be sucked into a maelstrom? Or where it will drift?' She stopped herself from thinking of the misfortunes that the future might hold for her, because she could see nothing but a profound darkness before her. Therefore the enthusiasm for reforming her life which had made her seek Bithal Das's assistance, had now cooled. Even if Bithal Das had promised her a hundred rupees a month at this moment she would probably have refused. But since she herself had made the suggestion, it embarrassed her to take it back at this stage. She said to Bithal Das, 'I want to think about it. And I will give you my answer tomorrow.'

Bithal Das: What do you need to consider?

Saman: Nothing much, but give me a day.

It was ten o'clock at night. The golden moonlight of winter was spread everywhere. Saman was watching the blue sky through her window. As the moonlight had dimmed the stars, so her heart's desire had eclipsed her pious resolution. She was trying to solve a difficult problem: What answer should she give to Bithal Das?

That morning she had put him off with the excuse that she needed time to think and would give him her answer the following day. That was an excuse but now, after a day spent in deliberation, her thoughts had again undergone a change.

Although her present circumstances provided Saman with a life of luxury, at times she was compelled to extend hospitality to people whom she hated, and whose conversation she found revolting. So far she had retained her finer feelings. They had not yet been substituted by a disposition—one that was base enough to allow desires and adornments to destroy all emotions. It is true that she loved adornment and elegance, but at the same time she detested the carnality that was needed in order to procure these luxuries. Sometimes when she was by herself, she used to compare her present life with her past existence. It is true that in the past such opulence as she now enjoyed was not within her reach. But on the other hand she was regarded with much respect within her close circle. Her neighbours had a high opinion of her. Living among them she was proud of her upper class parentage; and she impressed them with her piety. She was never ashamed before any of them. But here she was often mortified to think that even the lowliest of women would consider herself superior to her. The indignities she had to put up with in her past life were trivial compared to the humiliation she now had to suffer. The indecent remarks and innuendoes of some of her admirers cut her heart like daggers. At such moments her miserable heart would gnash its teeth with anger at Padam Singh, and she would think, 'If that cruel man had allowed me to stay on in his house for a couple of days more I might have gone back to my own house, or perhaps even Gujadhar would have sought a reconciliation and taken me home; and then life would have gone on as before.' This was the reason why she had asked Bithal Das to bring

Padam Singh with him—so that she could lacerate him with the arrows of her anguished heart.

But that day, when she realized how earnest he was in his desire to help her, her hatred gave way to respect, and she chided herself for blaming him for her own waywardness. 'He is a good man' she thought, 'He has regretted his haste in bidding me leave his house'. I will fall at his feet and tell him, "God will reward you for being so merciful to this wretched creature", I will return this bracelet to him too, so that he is satisfied that the woman on whom he is bestowing his help is not undeserving of it. And as soon as I have done this I will leave this dark realm of sin.

'But how can I forget Sadan?'

Saman was exasperated at her own weakness. For the sake of this short-lived attachment, the finale of which will be nothing but unfulfilled yearning, shall I give up the chance of mending my life? To bask in an evanescent moonlight, shall I spend the rest of my life in a dark pit? Moreover, shall I also ruin the life of a simple young man whom I love from the bottom of my heart? Such disloyalty to a person who has been so generous to me! No, I will pluck this love from my heart and hurl it away! I'll forget Sadan. I'll tell him "Forget me, and allow me to accept this helping hand that will enable me to put this life of shame behind me."

'Alas, I was deceived. How alluring, how pleasant this place had seemed from a distance. I took it for a blooming garden. But what has it turned out to be? A fearful wilderness, full of bloodthirsty beasts and poisonous insects! How beautiful this river looks from a distance— like a sheet of moonlight. But what does it harbour? Monstrous river creatures. It is also the playground of the sons of the rich!'

Saman was deep in these thoughts. She was impatient for daylight and the arrival of Bithal Das. She was desperate to be gone. It was midnight but she couldn't sleep. 'What if Bithal Das does not come tomorrow? Will I have to continue listening to the flattery of singers and instrumentalists from morning to night? Will I have to entertain these painted puppets for ever?' Saman had not lived more than six months in the red-light district. But she was already weary of it.

All day the musicians thronged her apartment. They used to relate tales of their own corruption and debauchery with great pride. There were sly pickpockets among them, clever gamblers, thieves, and thugs. And all of them were puffed up at their own capacity for shamelessness and mischief-making. Prostitutes who were her neighbours also visited her, dyed and decked out, shimmering like chandeliers. But they were golden vessels, full of fatal poison. They were shallow and tawdry, mean and treacherous, and hypocritical. They related stories of their own shamelessness and disgrace with great relish. They were forever eyeing the wealth of others, and conspiring to trap and dupe the gullible. In this society, upright citizens were abused and ridiculed freely. They were alluded to contemptuously. All day and half the night, the talk revolved around subjects such as thefts, robberies, adultery, murder, abortions, and embezzlement. And sometimes the protagonists themselves would divulge details of Bholi Bai's intimacy with the wealthy, the perception of which, on the day of Holi, had first awakened in Saman the feeling of her own helplessness, but which now showed her its true face. It was not what she had thought it was—it was not love, it was just licentiousness, merely ribaldry. Full of sensuality and devoid of finer feelings, it was pure lust. So far Saman had been putting up with all the unpleasantness, telling herself that if she was living in this abode of sin she couldn't run away from its atmosphere, that if she was resident in hell, she would have to be bound by the rules of its culture.

When Bithal Das had first come here she had been pointedly indifferent. That was because she had not yet become familiar with all the aspects of the life she was living. But by now she was wiser, and seeing the doors to deliverance open before her she was loth to spend another moment in the grip of this illicit way of life. As when, given the chance, the worst urges of a human being are awakened, so also when the opportunity presents itself the best impulses come to the fore. It was three o'clock at night. Saman was still restlessly tossing in bed. Again and again she thought of Sadan with longing. As morning drew near her agitation increased. She told herself 'Why do you exult in this love? Don't you know that its roots lie in the superficial attraction of good looks. This is not love, it is merely lust. Nobody comes to this place out of pure love. All they look for here is recreation. So don't fall into this trap. The prime of life is like a rising river. Everything that comes in its way is swept away by the gushing water. But when the mature river is flowing slowly through the plains what can it accumulate except insignificant grasses from its banks?

'If only I could win Sadan, and hold him with tenderness and devotion. I could entertain him and serve him, and keep his love for me fresh and evergreen. But in this place I cannot do so. Here, he will not trust me even if I give my life for him. The air in this place breathes suspicion. Yes, if I lived in a hut far from here, he could be mine. But that is impossible. Deliverance lies in running away from this place.'

When day broke, Saman fell asleep.

– 21 –

It was evening. Saman had waited for Bithal Das all day, but he had not arrived. Saman's misgivings were coming true. 'Perhaps he will never come. Some complication must have come up, or he may have got involved in some other business. Or perhaps the people who had promised assistance have let him down. But whatever the reason, he should have come here just once more to let me know what was decided. If nobody wants to help me, so be it. I will help myself. I need the support of one good man, though. Can't Bithal Das give me even that? Let me go and see him. I will tell him, "Don't worry, I don't need financial support. Only find me a place to live and some means of earning my bread. I will not ask for any more." But where does he live? I don't even know that, how can I find him? I may as well go to the park. People go there for a stroll. Perhaps I'll find him there. Padam Singh also goes there every day. If I meet him there I can give him the bracelet, and that will give me a good excuse to talk about this matter.'

Acting on her plan, Saman sent for a hired carriage and left in it, alone. For a long distance she gazed through the venetian blind but could see neither Padam Singh nor Bithal Das. She did see Sadan though, galloping on his horse, in her direction.

Saman's heart leapt up. It seemed to her that she had seen him after many years. Perhaps a change of locale has a refreshing effect on love. She was tempted to call out to him but stopped herself. And until he disappeared from her sight she continued to gaze at him with love. 'Such a handsome young man is devoted to me', this certainty bathed her heart in happiness. Sadan had never before appeared so good-looking to her.

The carriage drove on towards Queen's Park. This park was far from the city and attracted very few people. But Padam Singh's fondness for meditation brought him here regularly. He would sit on a bench in the wide, green grounds, lost in his thoughts. As soon as the carriage entered the precincts of the park, Saman saw him. Her heart began to tremble, and embarrassment and remorse began to pull her back. If, before leaving her house, she had known how she was going to feel, she would not have come here at all. But having ventured so far and seeing Padam Singh before her, it would have been senseless to go back with her mission incomplete. She descended from the carriage and summoning her courage, began to walk towards Padam Singh, the way voice travels in the opposite direction of the wind.

Padam Singh was looking at the carriage with astonishment. At first he did not recognize Saman. He was wondering who this woman, coming towards him, might be. At first he thought it was a Christian lady, but when Saman came closer he recognized her. He looked at her once with lowered eyes and then nervousness got the better of him. When she came and stood before him with her head bowed, he started throwing helpless glances this way and that, as though looking for a mousehole to hide himself in. Suddenly he stood up and turning around, began to walk away rapidly. Saman was stupefied. She had come with some hopes, and had certainly not envisioned this scene.

'How contemptible he finds me, to shun me so!' she thought. The devotion that she had begun to feel for him, vanished. She said, 'I have come to tell you something. Please wait, I will be much obliged.'

Padam Singh sped up his retreat, as one fleeing from a ghost. Saman could not take this rebuff. She said sharply, 'You need not be afraid, I have not come here to ask for anything. I came just to give you this bracelet. Here! Now I will leave.'

Saying this, Saman threw the bracelet in the direction of Padam Singh, who now hesitated, looked at the bracelet which lay on the ground before him, and recognized it. It was Sobhdra's bracelet. Saman had gone back several paces, in the direction of her carriage. Padam Singh now quickly overtook her and asked, 'Where did you find this bracelet?'

Saman replied with indifference, 'If I take no notice of what you have to say, and go away, you will have no right to complain.'

Padam Singh: Saman Bai, don't embarrass me. I feel I can't show my face to you.

Saman: Why?

Padam Singh: I think again and again that if on that day I had not asked you to leave my house, matters would not have come to this.

Saman: You should not reproach yourself. You did me a favour by turning me out. Your action exalted my life.

Stung by the sarcasm, Padam Singh replied, 'If it was a favour to have changed your life in this way I will not claim it and am not proud of it. It is Bithal Das and Gujadhar you are beholden to'.

Saman: You may disown your good work, but I still feel obliged to you. Padam Singh, don't make me spell it out. I know it was all there in my destiny, but I did not expect such disregard from a decent man like you. Perhaps you think that it is only the great who are greedy for esteem, but believe me, the lowly have even more need of it, because they have no way of finding it. To achieve it they cheat, deceive, and are dishonest. The happiness that comes from esteem is found neither in luxury nor in wealth. I was always haunted by this craving. How was I to find it? I found the answer to this question many times, but when I found it in your house on the day of Holi, all my misgivings vanished. If I had not gone to your function I might still have been living contentedly in my hovel. But I considered you a god— infallible—and your high spirits that day were my inspiration. Bholi Bai was sitting in splendour before you, and you were chivalry itself in your attitude to her. Your friends were dancing attendance on her. The scene had on me the effect it was bound to have on any simple woman hungering for appreciation and esteem. But why should I talk about this. Why should I blame you?

'It was all preordained.' Saman was going to continue, but Padam Singh who was listening to her with grave interest, interrupted her, 'Saman, are you saying all this to embarrass me or is it a fact?'

Saman: My intention was to embarrass you, but it is a fact all the same. I had erased all this from my memory a long time ago, and if you had not shown such indifference today I would not have brought it up. I already regret having said it, it is all so irrelevant now. Forgive me.

Padam Singh did not raise his head; he was lost in reflection. Saman had come with the intention of thanking him for his generosity, but the conversation had taken a turn that left her no opportunity to do so. It seemed to her that after so much bitterness, any expression of gratitude or acknowledgment of obligation would be ill-timed. She turned and went towards her carriage. Suddenly Padam Singh asked her, 'And this bracelet...?'

Saman: I saw it at a jeweller's yesterday. I recognized it because I had seen it on Sobhdra's wrist, so I picked it up.

Padam Singh: What did you have to pay for it?

Saman: Nothing. In fact he was intimidated.

Padam Singh: Can you tell me the name of the jeweller?

Saman: No. I've given him my word that I wouldn't.

Saman then left. Padam Singh remained sitting for some time, then he lay down on the bench. Every word that Saman had said rang in his ears. At that moment he was so lost in thought that even if somebody had come and stood before him he would not have noticed. It seemed as though he had suffered a harsh blow. He was a sensitive man. During a row, if Sobhdra ever said anything cutting to him, it would distress him for weeks. He was proud of his style and mode of life, his views, his sense of duty. Today that pride had been shattered. The guilt of which he had rid his own conscience, by passing it on to Gujadhar and Bithal Das, had come back with a hundredfold heavier burden, to rest on his head. He couldn't move his head under its weight, it was crushing him. Thoughts gave way to imagination, and imagination gave birth to hallucination. He heard a voice from far away that said, 'If I hadn't gone to your function I would have been living contentedly in my hovel.'

A wind started up. The leaves that covered the dark trees began to sway and rustle as though nodding admonishment, 'It's you who ruined her life, it's you...'

Padam Singh stood up in agitation. In front of him there was the tall spire of a church. The clock in the tower was chiming the hour. The chimes seemed to be saying in their melodious voice, 'It's you who ruined her life.'

Like a traveller in a forlorn place, who, seeing a large black cloud forming in the sky, hurries to take refuge under a single distant tree, Padam Singh began to take long strides in the direction of habitation. But how could he leave his imagination behind? Saman was following him, and sometimes she would overtake him and crossing his path would stop him in his track, saying, 'It's you who ruined my life.' She would overtake him, sometimes from one side and sometimes from the other, repeating her words every time. Padam Singh said nothing and did not change his pace, like the plucky man who does not run when chased by dogs.

Finally his perilous journey came to an end. Padam Singh reached home, where he went straight to his room. He got into his bed and pulled the sheet over his head. When Sobhdra insisted that he eat something, he fended her off by pretending that he had a headache. But he could not fend off Saman who sat in his heart all night, cursing him. 'You are very proud of your knowledge and your wisdom. But you play with fireworks near huts of straw. If you want to splash your wealth, do so in some lonely place, far from where people live. Why must you wring the heart of the unfortunate poor? Why must you devastate them?'

'If you have the money, go eat your fill. But look, there is a destitute orphan standing by you. Mind his gaze of longing! If you can't share your goodies with him, go elsewhere. Don't make him suffer the pangs of deprivation.'

Early the next morning Padam Singh arrived at the house of Bithal Das.

– 22 –

In the evening Sobhdra missed her bracelet. She ran to the bathroom because she remembered taking it off there and putting it on the shelf. When she couldn't find it there she was worried. She looked in every shelf and every cupboard of her room, everywhere in the kitchen. Not finding it, her anxiety increased. She searched every trunk, every corner of the house, as though she were looking for a lost needle; but to no avail. She questioned the maidservant who denied seeing it, swearing by her son. She called Jitan and asked him. He replied, 'Mistress, don't disgrace me in my old age. I've spent my life serving respectable people and was never tempted. Why should I suddenly change?' Disappointed, Sobhdra once again opened trunks and bundles of cloth, looked inside jars filled with flour and pulses, and felt with her hand inside water containers. Finally she lay down on her bed, exhausted. She had seen Sadan go into the bathroom and it occurred to her that he may have hidden it as a joke, but she did not have the courage to ask him. She thought she would ask her husband when he came home for his meal. As soon as he arrived she told him, but Padam Singh replied carelessly, 'Look carefully. It must be in the house. Who would take it?'

Sobhdra: I've looked everywhere in the house.

Padam Singh: Did you ask the servants?

Sobhdra: I've asked them. Both swear they haven't seen it. But I remember very well putting it on the shelf.

Padam Singh: Well then, I suppose it had wings, so it flew off.

Sobhdra: I don't suspect the servants.

Padam Singh: Who else could have taken it?

Sobhdra: Shall I ask Sadan? I saw him go into the room. Perhaps he hid it as a joke.

Padam Singh: That doesn't make sense. If he had hidden it he would have told you by now.

Sobhdra: There is no harm in asking him. Maybe he wants to return it when he finds I am really worried.

Padam Singh: Of course there is harm in it! If he doesn't know anything about it he will think you are accusing him of theft.

Sobhdra: He did go into the bathroom. I saw him myself.

Padam Singh: So, do you think he went there to pick up your bracelet? That makes no sense. Anyway, don't even think of asking him. He'll imagine you are accusing him. There is no question of his taking it, and even if he has, he will return it tomorrow, if not today. 'What is the hurry?'

Sobhdra: I do not have your forbearance. I am really very distressed. Let me ask him, it will make me feel better.

Padam Singh: No. Don't ask him.

Sobhdra was silent for the time being, but later in the day when the uncle and nephew sat down for their evening meal, she said to Sadan, 'Sadan, I can't find my bracelet. If you have hidden it please give it back. Don't tease me.'

Sadan went pale. It was his first theft and he had not yet learnt how to put a brave face on one. He was caught with the morsel in his mouth, which in his agitation he forgot to chew. He pretended not to hear. Padam Singh gave Sobhdra such an infuriated look that she did not dare open her mouth again. Sadan quickly swallowed a few more morsels and left the room. Padam Singh said to Sobhdra, 'You do everything that I forbid you to.'

Sobhdra: Didn't you see his expression? He has taken the bracelet. I know I am right.

Padam Singh: Where did you learn to read the face so well?

Sobhdra: His face proclaims it.

Padam Singh: All right, you may be right. But if he has taken it what objection can I have to it? Everything that's mine is his, whether he takes it with my permission or without.

Sobhdra said with vexation, 'You have a slavish mentality; you can continue to be a slave. I don't have such patience with someone who takes my things.'

The next evening when Padam Singh returned from his walk, he
threw the bracelet before his wife. Sobhdra ran to pick it up. She
asked with surprise, 'Where did you find it?'

Padam Singh: That's not your business.

Sobhdra: Didn't I say Sadan had hidden it? Wasn't I right?

Padam Singh: You are talking nonsense again. I found it in a
jeweller's shop. By accusing Sadan you have embarrassed both him
and yourself.

– 23 –

Bithal Das suspected that Saman found the sum of thirty rupees a
month unacceptable, and so she had avoided the issue by making
the excuse that she needed time to think over it. Wondering how to
find the rest of the money, he did not go to see her the next day.
Instead, he racked his brains for a way to procure the money.
Sometimes he thought of taking a deputation to other cities and
sometimes he contemplated organizing a play. If he had had his own
way he would have sentenced all the well-to-do people in the city to
transportation for life.

Kanwar Anarudh Singh was a generous and cultured citizen. But
Bithal Das retraced his steps after arriving at his doorstep just because
he sensed that the people there were drinking. He thought, 'Why would
someone who lives in the lap of luxury, bother to help me?' In his
view, at that point of time, the greatest good was to help him, and the
greatest sin was to oppose him. He was still puzzling over the dilemma
of whether or not to go to Saman, when Pandit Padam Singh arrived.
His eyes were red and his face was pale. He had the look of someone
who had stayed awake all night. Although Bithal Das himself was at
that moment the embodiment of anxiety and frustration, he melted as
soon as he saw Padam Singh's plight. Very warmly he shook hands
with him and said, 'My brother, you look worried. Is everything all
right with you?'

Padam Singh: Yes, thank you, I am well. I hadn't seen you for a
long time so I thought I should drop by. What have you decided about
Saman?

Bithal Das: That is my predicament. In a city as large as this I can't
manage to find enough funds to make up thirty rupees a month. I am
beginning to suspect that I do not have the skills of a good fund raiser.
I don't know how to win over others; I can only blame them. But my

shortcoming of the moment is that I have been able to arrange for only ten rupees per month, permanently. There are plenty of millionaires, but their hearts are stone. It's not just the wealthy, even Parbhakar Rao turned me down. His writing makes one think that his heart is a boundless sea of deep feeling. After the Holi function he conducted a smear campaign against you for months, but when I approached him today he said, "Am I the only one who owes a debt to this nation? I wield the pen and my contribution to my country is what I write. It is the wealthy who should contribute their wealth." I was stunned by what he said. He is building a new house these days, and has also bought shares in coal, but he managed to elude me. Other people are embarrassed, at least. This man actually put the blame on me!

Padam Singh: Are you sure that Saman will agree to go to the Widows' Ashram if she is given fifty rupees a month?

Bithal Das: I am certain of it. It is another matter that the committee might decide against admitting her.

Padam Singh: In that case I will put an end to your anxiety. I will pay fifty rupees a month as long as I live.

Bithal Das looked at him with a mixture of astonishment and gratitude. Relief blossomed on his face. He leaped up and embraced Padam Singh, saying, 'My brother, what you have just said makes me want to put my head on your feet and weep. You have vindicated the Hindu nation. And you have blackened the faces of all the millionaires. But how will you shoulder such a heavy burden?'

Padam Singh: God will find a way for me.

Bithal Das: Is your practice doing well these days?

Padam Singh: No. In fact it is doing very badly. I'll sell my carriage and horse, that will save thirty rupees a month. The electricity will go, that will be another ten rupees. Ten rupees more I'll find somehow.

Bithal Das: It makes me unhappy to load you with the whole burden. But the rich of this city are no help at all. You will have to rent a carriage to take you to court, once your own vehicle is sold, won't you?

Padam Singh: No, I won't need a rented carriage because my nephew has a horse. I'll go to the courts on the horse.

Bithal Das: Is your nephew the same person who is sometimes seen at the marketplace in the evenings?

Padam Singh: Perhaps.

Bithal Das: He resembles you. A pink and white complexion, good looking and well-built. The young man has large eyes?

Padam Singh: Yes, your description is accurate. It is the same person.

Bithal Das: Why don't you stop him from wandering around in the marketplace?

Padam Singh: How do I know where he goes? Perhaps he goes there sometimes. But he has a clean character so I have never worried on that account.

Bithal Das: You are making a serious mistake. He may have had a clean character, but these days his habits are not clean. I have seen him in a place where he should not be, not once but many times. I think he is in love with Saman.

Padam Singh was flabbergasted. He said, 'You have given me very bad news. He is the hope of my family. If his character is sullied I will be in serious trouble; I won't be able to show my face to my brother.' Saying this, Padam Singh's eyes filled with tears. He continued, 'Please talk to him. If my brother gets an inkling of this he will never see me again.'

Bithal Das: Don't worry, he will be made to mend his ways. I had no idea he was your boy. Now I'll pursue him. In any case if Saman leaves the place tomorrow he will himself come round.

Padam Singh: Even if Saman goes, the district will not be deserted. He will find someone else. What shall I do? Should I send him home?

Bithal Das: He'll never stay there. In the first place he will not go, and even if he does he'll return the next day. The addictions acquired in one's youth are bad. This is all the doing of Beauty's spell which has enveloped some of the best parts of the city. Instead of keeping temptations out of sight we adorn and display them. We dig pits for our simple, naive youth; we awaken their dormant emotions. Nobody knows when this vulgar practice began. Where there ought to be libraries and centres for improvement of character, we allow instead these beauty marts. Alas!

Padam Singh: You did launch a movement in this connection, didn't you?

Bithal Das: Yes, I did. But, like you, others showed merely a verbal interest. As for me, I have neither financial resources, nor much education nor any contacts. Who will listen to me? They think I am mad. There are so many well-educated, talented people in the city; they are all quite content. Nobody bothers to listen to me.

Padam Singh was at that moment like the proverbial hot iron, and Bithal Das's blow struck its mark. He said, 'If I can help you in any way, I am at your service. This is not a shallow promise like the earlier one.'

Bithal Das jumped for joy. He said, 'If you help me, the sky will be the limit of what we can achieve. But excuse me for saying this: your resolves are usually not very strong. You are eager to help now but by tomorrow your enthusiasm will subside. Such work needs a lot of persistence.'

Padam Singh replied ruefully, 'God willing, you will have no occasion to complain this time.'

Bithal Das: If you have the determination we are bound to succeed.

Padam Singh: I can neither speak nor write [effectively, as you probably expect]. I'll simply follow you, with my eyes closed.

Bithal Das: All that will come, but empathy is what is needed. Determination can create fortresses out of thin air. Your speeches will have such power that people will be spellbound. But strengthen yourself, because the condition is that you must not let failures discourage you.

Padam Singh: You will have to guide me.

Bithal Das: Let me tell you my objectives, so that there will be no misunderstanding. My first goal is to remove the dancing girls from the prominent places and highways of the city. My second goal is to destroy the contemptible practice of holding dance and music performances. Do you have any objections?

Padam Singh: None at all.

Bithal Das: Do you have the same views on dance as you used to?

Padam Singh: Having burned down one home, would I be playing the same game? I don't know what happened to me in those days. I am certain now that my function incited Saman to leave her home. But here I have another notion: after all we have spent our lives in cities, how come we have remained immune to such bad influences? Moreover, dances have always been a part of city life, but one doesn't see them produce such extravagant consequences. This seems to prove that in such matters a person's disposition, inclinations, and habits play the decisive role. Can your crusade change people's temperament?

Bithal Das: That is not our object at all. We merely want to reform the conditions that assist the baser nature of man; that's all. Some people are strong by nature, they need no additional vitamins. There are those who benefit from a good diet and strengthening drugs. And there are also those who will always remain helpless and weak, and their emaciated bodies will show no change even if they are made to

live in barrels of butter. We are concerned only with the second type—
those who can benefit from the use of vitamins and strength producing
drugs; and it is they who form the bulk of the population.

– 24 –

When Saman returned from the park she began to regret the
accusations she had levelled at Padam Singh. She thought, 'Why
did I speak so harshly to Padam Singh? He assisted me with such
generosity and I rewarded him by blaming him for my own stupidity!
After all, dance and music are everywhere. The rich, the poor, and
people of all ages enjoy such entertainment. If I, goaded by the
temptations of my own mind, jumped into the fire, how was it Padam
Singh's or anybody else's fault? Bithal Das went to see all the rich
men in the city. He must have met the Seths who come here, also. But
none of them offered him any encouragement. Why? Because they
don't want me to be freed from this living hell. My departure would
interfere with their life of pleasure. Like cruel hunters they want to
wound me for the pleasure of watching me writhe in pain. There was
just one courageous man who stretched out his hand to pull me out of
this dark pit, and I humiliated him! How ungrateful he must have
thought me! How he ran when he saw me! I should have been the one
to be overcome with shame, but just for his avoidance of me I began
to insult him. When the hunter finds that a bird has avoided his trap,
how angry he becomes! When a boy happens to touch something
unclean, he runs around touching other boys in order to taint them as
well. Am I like that cruel hunter and the foolish boy?'

Ask an author how he feels about the piracy of a discriminating
plagiarist when compared to the applause of an undiscriminating
populace. At that moment, Padam Singh's avoidance of her seemed
more attractive to Saman than all the flattery of her admirers.

All night she was thinking along these lines. She made up her mind
that she would go to Bithal Das the next morning and tell him to find
a sanctuary for her. She would tell him, 'I am not asking you for
anything. It's just shelter that I want. I'll earn my living by grinding
grain and doing some sewing.'

Early the next morning, while she was getting ready to go, Bithal
Das himself arrived. Delighted, Saman said to him, 'I waited for you
all day yesterday. And now I was coming to see you myself.'

Bithal Das: Yesterday I couldn't come for several reasons.

Saman: So, have you found a place for me to live?

Bithal Das: I couldn't accomplish that, but Padam Singh has saved the day. He came to me a little while ago and promised to contribute fifty rupees a month.

Saman's eyes filled with tears. Padam Singh's generous sacrifice had made her heart overflow with feelings of devotion and pure love. She was full of remorse for her harsh words to him. She said, 'Padam Singh is a river of compassion and generosity. May God bless him with eternal happiness. I will always remain grateful to him. But the stipulations for leaving this place which I had made to you were merely a test to see whether you really wanted to help me. Now it is clear to me that both of you are angels in human form. I do not want to give you any more trouble. I was hungry for sympathy and support and I have been given both. I will find my livelihood myself, but I will trouble you to find me a place where I can live in safety.'

Bithal Das was astonished. His eyes brightened with pride. It was only in this sacred land that the thoughts of a mere dancing girl could have such depth! He said, 'Saman, I can't tell you how delighted I am to hear such noble words from you. But how will you survive without money?'

Saman: I will work with my hands. There are thousands of unfortunate women in this country who receive help from none but God. I will not make you pay for my shamelessness.

Bithal Das: Will you be able to stand the hardship?

Saman: I will endure everything. This place has taught me that shamelessness is more heartrending than hardships. Physical hardships hurt the body, but shamelessness murders the soul. I am grateful to God that he sent the two of you to help me in my need.

Bithal Das: Saman, at this moment it is you who are the goddess.

Saman: When can I leave this place?

Bithal Das: Today. I have not yet requested the committee to admit you, but never mind, you can go and stay there. If someone in the committee makes an objection, we'll deal with it later. But remember that you should not tell them anything about yourself. If you do, there will be an uproar among the widows.

Saman: I will do whatever you think is appropriate.

Bithal Das: We'll go this evening.

A little while after Bithal Das left, two women of the profession came to see Saman. She let them know that she had a headache and couldn't see them. Saman used to cook her meals herself. The threads

of custom are stronger than the chains of courtesy. She had decided to
fast that day. Prisoners don't want to eat on the day of their freedom.

In the afternoon a throng of noisy musicians arrived. Saman sent
them away with some excuse. She detested them and could not bear to
see their faces anymore. Towards evening Balbahadar Das sent a basket
of Nagpur oranges. Saman returned them at once. At four o'clock
Chaman Lal sent his phaeton for Saman's outing. She sent it back too,
immediately.

The break of day is a time when all creatures are in good spirits.
Birds hop in the trees, calves frolic. The end of Saman's long night of
despair found her in the same mood. She sent for a packet of cigarettes,
and a bottle of varnish which she kept on the shelf. She broke one leg
of a chair and stood the chair against a wall. It was barely five o'clock
when Munshi Abul Wafa arrived. This gentleman was very fond of
cigarettes. Saman welcomed him more warmly than usual and after
some small talk she said to him, 'I can offer you a cigarette that you
will never forget.'

Abul Wafa: I can't wait to smoke it!

Saman: Look, I sent for it from a British store, especially for you.

Abul Wafa: Then I will count myself among the blessed! Hurrah
for me, and hurrah for the power of my heart's desire!

Abul Wafa took the cigarette in his mouth, and Saman lighted a
match. Abul Wafa held out the cigarette in his mouth to be lighted,
but for some reason, instead of lighting his cigarette the flame lighted
his piebald beard. Like dry grass, more than half of the beard burned
down in the twinkling of an eye. He threw away the cigarette and
began to rub his beard with both hands. The fire was extinguished but
the hair was gone. He leaped up to survey the damage in the mirror.
The remnants of burnt hair were looking like the fibres of boiled corn.

Saman said with feigned repentance, 'Oh, what have these hands
done! I was only lighting the match.'

She was trying to suppress her laughter but did not succeed. He
looked so disconcerted, and there was such helplessness in his
expression that it seemed as though he was the most unfortunate being
in the world. Saman's ill-disguised mirth was the last straw. 'What did
I do to deserve such spite?' he managed to say amid anger and
humiliation.

Saman replied in a tone of remorse, 'Munshiji, may both my eyes
be blinded if I did this on purpose. Even if I owed you a grudge, the
poor beard had done me no harm.'

Abul Wafa: One enjoys the pleasantries of a beloved, but not if they burn one's face! It would have been better if you had burned my whole body. Where can I go with this charred face?

Saman: What can I do, I am full of remorse. If I had a beard I would have offered it to you immediately. But I believe artificial beards are available?

Abul Wafa: Don't make fun of me! If somebody else had done this I would have taught him a lesson.

Saman: Why are you making such a fuss? All you have lost is a few hair. They'll grow again in a couple of months.

Abul Wafa: Don't make me angrier. I might say something insulting!

Saman: You've lost all self control—just for a beard! Let's say I burned it on purpose. So what? You set fire to my virtue every day. Is my virtue worth less than your beard? It's not so easy to be a lover. Go home, I've no use for such selfish people!

Abul Wafa gave Saman a look of fury, then, taking a handkerchief out of his pocket he hid his burnt beard with it and left in silence. This was the same gentleman who felt no shame in exchanging pleasantries with a dancing girl in the market-place.

Now it was time for Sadan to come. Saman's desire to see him that day was intense. She was thinking, 'This will be our last meeting. Today this story of love will come to an end. I will never see his dear face again. My eyes will thirst for him but their craving will never be satisfied. I will never hear again those declarations of love and longing, drenched in genuine feeling. Illicit perhaps, but it was an honest bond. Oh God! give me the strength to bear this loss. It would be better if Sadan didn't come today. His not coming today would be good for me. My resolve may not be able to stand the test of his presence. But if he came I would talk to him frankly for once. I would try to stop him from sinking into despair.' Suddenly Saman saw Bithal Das stepping out of a rented carriage. Her heart pounded inside her chest. In a moment Bithal Das had come up. He said with surprise, 'You are not ready?'

Saman: I am ready.

Bithal Das: You haven't even tied your bedding.

Saman: I am leaving everything behind. This is my new birth.

Bithal Das: What about all this stuff?

Saman: Please sell it and use the proceeds for one of your charities.

Bithal Das: All right. I'll lock this place. Now come down to the carriage.

Saman: I cannot go before ten o'clock. I have to take leave of my lovers. Go and sit on the roof. I am ready otherwise.

Bithal Das didn't like that but stopped himself from voicing his disapproval. He went up to the roof obediently and began to stroll.

It was seven o'clock but Sadan did not arrive. Saman waited for him until it was eight, then she lost hope. It was the first occasion since he had begun to visit her that he had neglected to come. It seemed to Saman that she was lost in a desolate landscape. Her heart was overflowing with a stinging-sweet, painful-delicious desire. Again and again she asked herself, 'Why hasn't he come?' Then she thought perhaps he had had an accident, and this thought made her tremble.

At eight o'clock, Seth Chaman Lal arrived. As soon as Saman saw his carriage she went and sat on the balcony. Sethji climbed the stairs with difficulty and demanded in a breathless voice, 'Where is the lady? Why did she return the phaeton today? Have I offended her?'

Saman: Come out on the balcony. It's hot inside. I had a headache today and didn't want to go out.

Chaman Lal: Why didn't you send Hira to me? I could have sent you a medicine from the hakim. He knows some good preparations.

Saying this, Chaman Lal sat down on the chair. But the chair with one missing leg toppled over, and the Seth's legs went up while his head went down. He lay in a heap on the ground like a mound of grass. Just the one exclamation, 'Aray!' escaped his lips and then he was silent. Flesh had overcome spirit.

Saman was struck by the fear that he had been badly hurt, but when she saw him by the light of the lantern she couldn't help laughing. The Seth was lying as motionless as though he were buried under a mountain. He was mumbling 'Oh, oh, my back is broken. Call my groom, I want to go home.'

Saman: Are you hurt badly? You should not have pulled the chair away from the wall. I'm sorry I forgot to warn you. But you are not even trying to pull yourself up. Just lying on the ground like this!

Chaman Lal: My back is broken and you find it funny!

Saman: So how is it my fault? Had you been lighter I could have helped you sit up. Now you have to pull yourself up. Try. You can do it.

Chaman Lal: I don't know how I'll get home. I must have left home at an inauspicious moment. Walking down your stairs will be hell. Why did you do this to me?

Saman: I am very sorry about it.

Chaman Lal: Don't try to bluff. I know you did it on purpose.

Saman: Even if I owed you a grudge, what could I have against your poor back?

Chaman Lal: I'll be cursed if I come here again!

Saman: Sethji, you are losing your temper for nothing. Even if I did it on purpose, can't you put up with a little teasing from your beloved?

Bithal Das now came down from the roof. Chaman Lal turned pale when he saw him. His humiliation was complete.

Bithal Das controlled his mirth and asked Chaman Lal 'How come you are here Sethji? I must say I am astonished to see you here.'

Chaman Lal: Don't ask me that now, brother. I'll be damned if I ever come here again! Just help me downstairs.

Bithal Das held one of his arms, Saman held the other, the groom held his back, and so they brought him downstairs and put him in his carriage.

Upstairs, Bithal Das told Saman, 'The carriage is still waiting. It is ten o'clock. We should leave now.'

Saman: There is just one more thing to be done. Pandit Dina Nath should be coming soon. Let me just deal with him. If you could step upstairs again, for a moment...'

As soon as Bithal Das went upstairs Pandit Dina Nath arrived. On his head was a banarsi safa* and he was clad in an achkan and a thin dhoti* with a black border. On his feet were highly polished black pumps. This attire set off his fair skin to advantage.

Saman: Welcome, sir. Let me touch your feet.

Dina Nath: Bless you. May your youth be prolonged. May you trap many wealthy fools. May your fortunes improve day and night.

Saman: Why didn't you come yesterday? I kept the others waiting for you and sat up half the night.

Dina Nath: I got caught in this affair. Dr Shyama Charan and Parbhakar Rao forced me to go to a forum for independence.* They were all babbling there. They told me, 'Why don't you say something?' But I said, 'Do you take me for a fool?' and somehow I got away.

Saman: Many days ago I had asked you to have my shutters varnished, but you told me that varnish was not available. I sent for a bottle today. Please have it applied tomorrow.

Pandit Dina Nath was sitting directly under the shelf where Saman had kept the varnish. Saman picked up the bottle but somehow its bottom came off and... Panditji was covered in varnish. It seemed as if he had fallen into a trough of syrup. He sprang up and taking off his safa, began to wipe himself with a handkerchief. Saman said, 'I wonder if the bottle was broken. A pity all the varnish is gone.'

Dina Nath: You are worried about your varnish when all my clothes are soaked and I don't know how I'll get home.

Saman: Who can see you in the dark? Just go away quietly.

Dina Nath: Stop advising me! My clothes are ruined and you change the subject to divert me.

Saman: Do you think I dropped it on purpose?

Dina Nath: Can I read your mind?

Saman: All right, then I dropped it on purpose.

Dina Nath: I am not blaming you. Drop some more if you like.

Saman: You can charge me for your clothes.

Dina Nath: Don't be so angry. I am not reproaching you.

Saman: You talk as though you are doing me a favour.

Dina Nath: Saman, don't embarrass me.

Saman: You lost your temper just because your clothes were spoilt. This is the love that you never tire of professing! Now I know its true colour. I've been warned in time! Now go home and don't come back here again. I don't need windbags like you.

Bithal Das was watching this drama from the roof. He understood that the show had ended, and came down. When he saw him, Dina Nath gave a start and picking up his stick, speedily went down the stairs.

After a little while Saman came downstairs. She was wearing a simple white saree and there were no bangles on her wrists. Her look was sad, though not because she was leaving behind a luxurious lifestyle but because she regretted having fallen in this fiery cave in the first place. There was no wistfulness in her sadness, only courage. This was not the pallor of the drunkard's face, but the pallor that can be seen on the faces of brides and bridegrooms—in which are hidden their good resolutions and anxieties about their coming responsibilities.

Bithal Das secured the door with a padlock and sat on the coach box. The carriage started on its journey.

The shops in the marketplace were closed but the road was not. Saman looked out of the carriage window and saw a row of lighted lanterns. As the carriage advanced, so did the row. One lantern followed another at some distance, but the whole bright, glowing row kept running far away like her desires. The carriage was moving fast, the boat of Saman's life was also moving fast through the sea of thought, swaying a bit, and getting caught now and then in the golden net of stars.

– 25 –

In the morning when Sadan went into the women's quarters, he saw the bracelet on his aunt's wrist. Shame overpowered him and he could not lift his eyes with embarrassment. He finished his breakfast quickly and left the room, wondering where she had found the bracelet.

'Was it likely that Saman had sent it here? Did she know whose it was? I didn't even tell her my address. Perhaps this is a second bracelet of the same design. But it was hardly possible to make it so soon. Definitely, Saman had found out about it and returned it to Aunt.'

Sadan thought about this for a long time but each time he came to the same conclusion. 'Even supposing Saman found out my address, was it appropriate for her to send my gift here? It was a kind of treachery, in fact. If she does know my circumstances now, she must think me a sly and deceitful individual. That she sent the bracelet to Aunt proves that she thinks I am a thief as well.'

That evening Sadan could not summon up the courage to visit Saman. If she thought him a thief and a fraud, he was not inclined to meet her. Sadan was very dejected. It depressed him more to stay at home, but he still did not feel like going anywhere. Unwillingly, he resolved to stay in, and lay down on the bed. He had decided against going to see Saman.

A week passed in this way. He was longing to see Saman and his desire was burying his suspicions. In the evenings he would feel much worse, like a person who is depressed after an illness and does not want to speak to anyone; for whom the least physical activity becomes a chore so that he does not leave the place where he finds himself. That was the state Sadan was in.

Finally, by the eighth day, Sadan could bear it no longer. He sent for his horse and rode out to see her. He had resolved to tell her everything. He thought, 'How can one hide anything from the person one loves? I'll fold my hands and tell her, "With all my faults it's you I love. You can punish me for my sins. If I stole or if I lied, it was you I did it for, so forgive me".'

Restlessly, he left the house at five o'clock. He went first to the river, where the cool, pleasant breezes calmed him. The fish flying out of the clear blue water and its golden currents reminded him of the playful looks of a maiden darting through a fine veil.

Sadan dismounted from his horse and sitting on the bank of the river, began to enjoy the enchanting landscape. Suddenly he saw a sadhu with matted hair and red eyes emerge from behind some trees. He was wearing a necklace of beads and instead of an awesome mystic air he had an expression of gentleness and simplicity on his face. Seeing him come close, Sadan stood up respectfully.

The sadhu grasped his hand informally as though he had always known him. He said, 'Sadan, I have wanted to see you for a long time, to tell you something that is good for you. You should stop going to see Saman Bai, otherwise you will be ruined. You don't know who she is. In the blindness of your love you can't see her faults. You think she is infatuated by you, but you are foolish. How can a woman who deceived her husband be consistent in love? You are probably going there now, but listen to a sadhu and return home. Your safety lies in doing so.' Having said this the sadhu went back in the direction he had come from and was out of sight before Sadan could speak to him.

Sadan began to think, 'Who was this sadhu? How did he know me? How does he know my secret?'

A number of factors: the silence of his surroundings, the agitation in his heart, the sudden appearance of the holy man, and the revelation of his secrets, all combined to make the sadhu's speech seem like a voice from the Unseen. Sadan's heart trembled with fear at the thought of an approaching calamity. He no longer had the courage to go to Saman, so he mounted his horse and thinking about the astonishing episode, went home.

Ever since Sobhdra had voiced her suspicion of Sadan in connection with her bracelet, Padam Singh had been angry with his wife. He was grieved by her meanness. He did not go into the women's quarters much as a result, and this upset Sobhdra. Padam Singh wanted to send Sadan away, while Sadan too now wanted to leave as there was no longer anything to hold his interest; but neither could voice their thoughts. It so happened however, that the very next day a letter arrived from Pandit Madan Singh which satisfied everybody. It informed them that Sadan's marriage was being arranged, and Sobhdra was asked to accompany Sadan to the family home.

Sobhdra was happy at the news. She thought she would enjoy the change for a month or two. There would be singing and merry making, and she was tired of her isolation, living as she was in a sort of a cage; so she could not hide her delight. Seeing her good spirits Padam Singh was even more dejected. He said to himself, 'In her joy she cares nothing for me. We won't see each other for months but she looks as happy as though she had found a treasure.'

Sadan also made preparations to leave. Padam Singh had thought that he would make some excuse to stay back, but this did not happen.

It was eight o'clock. The train was to leave at two, so Padam Singh did not go to court. He went into the inner rooms several times but Sobhdra had no time to talk to him. She was busy with her jewellery, her clothes, and her hair. The jewellery was being cleaned, the paandan* was being polished. Several women from the neighbourhood were visiting. In her excitement, Sobhdra had no appetite, though she had made the puris and sent them to the outer rooms for Padam Singh and Sadan.

Finally it was one o'clock and Jitan brought the carriage to the door. Sadan brought out his trunk and bedding and put it in the carriage. It was then that Sobhdra thought of Padam Singh and she asked the maid to look for him. The maid went to the outer rooms but he was not there, she searched for him downstairs, but could not find him there, either. Aware of her husband's sensitive temperament, Sobhdra guessed what had happened. She announced, 'I will not go unless he comes.'

Padam Singh had not gone out; he was sitting on the roof of the house. When it was one o'clock, and Sobhdra did not come out of the house, he came downstairs in irritation and said, 'Why are you still here? It is past one o'clock.'

Sobhdra's eyes filled with tears. Padam Singh's harshness, just when she was about to leave, had wounded her. Padam Singh also regretted his unfeeling tone. He wiped her tears, embraced her, and brought her to the carriage. When they arrived at the station, the train was ready to leave. Sadan jumped into the carriage. As soon as Sobhdra got in, the train started moving. She stood at the window and gazed at her husband until he was out of sight. By the time the lights were lit the train had arrived in Chinar. Pandit Madan Singh was at the station.

Sadan ran to greet him, he had never been so respectful to his father before.

On the journey from the station to the village, Sadan was getting more and more excited. 'At the wrestling arena they must be getting ready for the bout. They will be startled when they see me. People will come from all around, Mother will come running!' When there was only half a mile more to go and the horse was slow in moving through the rice fields, Sadan dismounted, and handing over the reins to the groom, he walked quickly towards his home. But when he arrived at the village there was silence all around. It was not even eight o'clock and people had closed their doors. Sadan came home and bent his head over his mother's feet. Bhama blessed and embraced him. Then she asked, 'Where is she?'

Sadan: She is coming. I came through the fields.

Bhama: So have you had enough of your uncle and aunt?

Sadan: What makes you say that?

Bhama: Your face tells it all.

Sadan: I don't think so! I am stouter than before.

Bhama: Liar! Your aunt must have starved you.

Sadan: Aunt is not like that. She looked after me very well.

Bhama: Then why did you ask for money?

Sadan: I was testing your love. I took just twenty-five rupees from you during all this time. But I got seven hundred from Uncle. The horse I bought cost four hundred. Then there were silk garments, and I strutted about the city as a rich man. Aunt used to make fresh halwa early in the morning and I had a seer of milk every day. In the afternoon there were sweetmeats and fruit. Will I ever forget how comfortable I was! I knew that once I started earning I could never be so well-off, so I thought why not make the most of my time there. I did everything I ever wanted to do. Uncle is a god. I had no idea he loved me so much! He never refused anything that I asked for.

It seemed to Bhama that Sadan's style of talking had become somewhat urban and unfamiliar. But her mistrust of Padam Singh and Sobhdra had been allayed. The next day there was a gathering of prominent villagers and Sadan's engagement ceremony was performed. This was the first link of the iron chain which would tie his feet, but Sadan was so consumed with his desire for love that even the thought of the bondage to come, could not make him demur. His love for Saman was not of the kind in which passion and the yearning of the soul are one. In his love for her, passion predominated. In spite of permeating his heart, Saman had not become an inalienable part of it. If he had possessed an inexhaustible treasure he would have devoted it to her, he could have sacrificed for her all the happiness that his life might bring, but he could not think of making her a partner in his sorrows, or his failures. He could share luxury with her but not misfortune. In Saman he did not feel the perfect trust which is the essence of love. He felt now that he would find in his marriage something that he could never find with Saman. He would see, and show, the real face of love. He would no longer have to make a pretence of love.

Such thoughts made Sadan impatient for his new love. His only misgiving was, 'What if she is not beautiful?' Beauty of face is a natural blessing which is unchangeable. On the other hand beauty of character is an acquired attribute which can be developed with education. Sadan investigated the matter of his betrothed's looks with the help of his close friends, who made enquiries from his in-laws' barber.* The barber's description of the girl gave her a resemblance to Saman. Sadan was well-satisfied, and began to make preparations to receive the new bride.

– End of Part One –

PART TWO

It is probably not correct to say that God always uses rational and obvious means to provide food to his creatures. Pandit Oma Nath enjoyed all the blessings of this world, without any obvious means. He owned no buffaloes or cows but there was no shortage of milk for his family. He did not till the land but there were stacks of grain in his house. If anywhere in the village, fish was caught, a goat killed, mangoes gathered, some festivity took place, his share would arrive, unsolicited. Amola was a big village with a population of three thousand people, but nothing ever happened in the whole village without his help and advice. If women wanted to order jewellery, they would ask for his help, marriages were arranged through him, documents for mortgages and earnest money were prepared under his direction, his counsel and help were sought in initiating legal activity. But, strangely enough, his authority and influence did not stem from either artfulness or courtesy. He was dry and short with the villagers, and was known for his harshness of tongue. Yet, the people put up cheerfully with his harshness. What was the secret of his personality, nobody knew. Some said it was magic while others attributed it to his noble destiny. However it is our view that it was the result of his perceptiveness. He knew when he needed to bend and when to stiffen. He felt it was in his interest to be firm with the villagers and to bend to those in command. All the government office holders in the district, from the highest to the lowliest, were on good terms with him. He could apprise an official of his approaching promotion, while petty revenue officials, bailiffs, and accountants were uninvited guests at his door. He would write an amulet for one, and give a charm to another. And to those who were undeceived by such pious fraud, he gave pickles and preserves.

The police superintendent considered him his right-hand man. Where he could make no headway, his help could procure him the utmost profit. So, there was no reason for the villagers not to worship such a resourceful man.

Oma Nath loved his sister Gangajali. But very soon after her arrival at her parental home, Gangajali discovered that her brother's love could not make up for the coldness of her sister-in-law. Oma Nath deeply regretted bringing his sister to his house. To please his wife he would agree with whatever she said: What right did Gangajali have to wear clean clothes, now that her husband was behind bars? No matter how pampered Shanta had been earlier, what right had she to expect to be treated as an equal of Oma Nath's own daughters? Oma Nath listened to his wife's malicious tirades and concurred. Gangajali would take her frustration out on her brother. She felt that by siding with his wife he was encouraging her sister-in-law to mistreat her. She felt that if her brother rebuked his wife sharply she would not dare to persecute her. Whenever he got a chance, Oma Nath explained his position to his sister, though even this was done with difficulty as, first of all, Janhvi made sure that he was not able to speak privately to his sister, and secondly because Gangajali herself did not trust a sympathy which seemed to make no difference.

A year passed in this way. Gangajali fell ill from grief and anxiety and developed a fever. At first Oma Nath brought her ordinary medicines but when they made no difference he became worried. One day while Janhvi was visiting a neighbour he went into his sister's room. She was lying unconscious on a bedding that was in rags. Her saree was tattered as well. Shanta was sitting beside her, fanning her mother. Oma Nath wept when he saw this miserable scene. It occurred to him that this was the very woman who used to have two personal maids. He was full of remorse for his own indifference. Sitting by her bedside, he wept as he said, 'I did you harm by bringing you here. I did not know it would come to this. I'll go and get a ved* today and God willing you will soon recover.'

Meanwhile Janhvi returned. She had heard his last words and began to say, 'Yes, yes, run and get the ved. If you don't, the heavens will fall. Only recently I had fever for months, but you did not run to the ved. If I had taken to my bed like her, you would have realized that I was ill. But how could I stay in bed? Who would have done the housework then? Such luxury is not my lot.'

Oma Nath immediately lost courage. He did not dare to call the ved, although he knew that if he did not call the ved, Gangajali might die in a few days instead of a few months.

Gangajali's condition deteriorated every day, until she developed a condition known as flux in which she passed copious amounts of blood. There was no more hope left for her. How could her liver which could not tolerate soft sago digest bread made of barley? Her weak body could no longer bear more torment, and suddenly one day death devoured the poor sufferer.

Shanta now no longer had anybody to turn to. She wrote two letters to Saman but received no reply. Shanta thought, 'My sister has forgotten me. Who cares for those that are hit by bad luck?' When Gangajali was alive Shanta would cry with her head in her mother's lap. Now she did not even have that support. Now she could press her head against the walls of her room to have her cry, but the difference between a mother's lap and hard walls is great. One is like a river whose water is rippled by soft breezes, the other is a desert full of fiery dust.

Shanta knew no peace now. Her heart burned like fire. She believed that her uncle and aunt were her mother's murderers. When Gangajali was alive Shanta used to shield her from her aunt's harsh words. She would run to obey a command at the slightest hint from Janhvi, lest she say something to hurt her mother. One day Gangajali dropped and broke the pot that held the ghee. Shanta told her aunt, 'It dropped from my hands.' She put up with the abuse that followed because she knew that her mother's heart could not bear Janhvi's taunts.

But after her mother's death Shanta was no longer afraid. Her helplessness had made her bold. She had much less fortitude now and was more prone to anger. She would often answer back when scolded. She had prepared herself against the harshest treatment. She was in awe of her uncle but her aunt did not intimidate her at all anymore. As for her cousins, she answered them back all the time. She was like the cow who, terrified by her bloodthirsty pursuers, grazes rapaciously in other people's fields.

Soon a year had passed. Oma Nath was trying hard to arrange a marriage for Shanta. But since he was hunting for an inexpensive transaction, his quest was not successful. He had raised two hundred rupees from the district and administration funds, but how could he find a bridegroom for so little? Had Janhvi had her way she would have got rid of Shanta by marrying her off to a beggar. But for the first time in his married life Oma Nath opposed her, and continued to seek an appropriate husband for his niece. Having sacrificed his sister to his timidity, he had now acquired some guts.

– 2 –

Even schemes for public welfare are dependent upon the inclusion of prominent people in the working committee. Although Bithal Das had no dearth of followers, his following consisted mainly of common people. The elite remained aloof from his projects. Padam Singh's participation put life into the movement. The slim trickle turned into a torrent. Important people began to take notice of it and trust it.

Padam Singh was not for long the only prominent person in Bithal Das's group. Even when we know that a project is good, we do not participate in it because we are afraid of being made a laughing-stock. We wait for an important person to lead the way, and when he does, we pick up courage. We are then no longer afraid of being singled out for ridicule. By ourselves we are afraid even in our homes; if we have just one companion we are unafraid even in a forest. Professor Romesh Dutt, Lala Bhagat Ram and Mr Rustum Bhai used to help Bithal Das secretly. Now they did so in the open. The list of influential supporters began to grow longer.

Bithal Das did not believe in being soft-spoken where the goal was to achieve cultural reform. For this reason people used to find him offensive—those who are sleeping comfortably find a sharp voice jarring. But Bithal Das did not care for their opinion of him.

Padam Singh was a determined person. He launched the campaign for the eviction of prostitutes with much enthusiasm. Bithal Das had some sympathizers among the members of the municipality, but they did not have the courage to put the scheme into practice. The matter was so difficult and complicated that people despaired of it. They thought that if the scheme were discussed there would erupt a crisis in the city. There were so many among the city's affluent, and businessmen, who had close connections with the red-light district. There were admirers, clients, jewellers, pimps. Who could dare to cross so many? Members of the municipality were mere puppets in their hands.

Padam Singh met the members and drew their attention to the problem. Parbhakar Rao's spellbinding articles were specially persuasive. Pamphlets were brought out and a series of speeches were made to awaken the interest of the people. Romesh Dutt and Padam Singh were practised speech-makers, so they took over this responsibility. Soon the scheme became an organized movement.

Padam Singh had achieved a breakthrough, but the more he thought about the campaign the more he despaired. He could not see how eviction could bring about the desired results. There could even be disadvantages that would outweigh the benefits. He felt that the best remedy for immorality was a high moral sense, and that there was no other solution for its eradication. Sometimes, when he thought about it at length he felt a certain dread, and apprehensions which he could not express, being a member of the movement. Although he had no hesitation in talking about public welfare before the common people, he was very reluctant to do so in the company of his friends. This became a severe test for him. Some would say, 'Why have you got yourself involved in this mess? Bithal Das has duped you! Why don't you enjoy life instead of taking on such useless projects.' Others would say, 'You must have been jilted by one of those beauties, that's why you are pursuing them with such determination!' To talk of reform before such friends was to invite ridicule. Even during his speeches, if he tried to sway the audience and play on their emotions, he would not find appropriate words. So far he himself was unmoved, and when he thought about this lack of feeling in himself, he knew that his heart was still uninvolved in his mission.

At the conclusion of a speech Padam Singh was more eager to find out whether it was a good speech, rather than whether the audience were affected and convinced by it. Yet, in spite of this attitude, the movement continued to grow. For Padam Singh such success was no less than a deep conviction of the validity of his cause.

There were still two months to go before Sadan's wedding. Free from domestic involvement, Padam Singh devoted himself to the campaign. He had lost interest in his work at court; instead, even at work the same subject was discussed. Gradually, his constant preoccupation fanned the sparks which finally grew into a strong love for his mission.

But as the date for the wedding drew near, Padam Singh began to feel a weakness in his knees. It occurred to him that he would be expected to provide entertainment for the wedding party, and tradition dictated that a performance by dancing girls was mandatory on such occasions. 'What would be my duty under such circumstances?' he asked himself, 'I should stop my brother from following such an odious custom, but will I be successful in doing so?' Delivering a speech on principles or morals to one's elders seemed out of place. Besides, 'Brother would be full of ambitious plans for his son's wedding. If he is disappointed he will be very unhappy. Nevertheless, it is my duty to adhere to my own principles.'

Even though Padam Singh was fully aware that there were very few people who would value his principled stand, and that most people would be sure to oppose him, he felt that it would be abject weakness to surrender his principles to popular demands, and so he decided not to hold the performance. 'If I am not able to institute a change for the better in my own household, it would be stupidity to attempt to reform others', he told himself.

Having resolved on his course of action, Padam Singh began to make preparations for the function. He considered it altogether inappropriate to spare any expense on such happy occasions. Also, he intended to make the arrangements so elegant and lavish that the absence of the traditional dance performance would not be attributed to niggardliness. One day, Bithal Das happened to see the sumptuous preparations. He was taken aback, and asked Padam Singh, 'What did you spend on this?'

Padam Singh: I will tell you when I get back.

Bithal Das: It can't be less than two thousand rupees.

Padam Singh: Yes, perhaps even a little more.

Bithal Das: You've thrown away this money. If you had spent it on some welfare work people would have benefited from it. If enlightened people like you consider such extravagance justified, what can be expected of the others?

Padam Singh: I don't agree with you there. If God has blessed somebody with the wherewithal, he should spend lavishly on such happy occasions. I don't say that he should put himself in debt, nor that he should sell his house for it, but he should spend whatever he can afford. When else can one indulge oneself, if not at such a time?'

Bithal Das: Do you think Dr Shyama Charan can afford to spend five or ten thousand rupees, or not?

Padam Singh: He can afford much more than that.

Bithal Das: But recently, at the wedding of his eldest son, he showed such restraint. There was no dance, no music, singing, no performances of any kind.

Padam Singh: Yes, he was very frugal over such worthless customs, but he made up for it by the lavishness of his banquets. So, what was the result of his economy? The people did not benefit from it. In fact, the money that could have gone to poor musicians, fireworks sellers, flower sellers, carpet spreaders, went instead, to rich firms. I don't call this frugality, I call it depriving the poor of their right.

– 3 –

It was nine o'clock at night. Padam Singh was sitting with his brother, discussing the arrangements for the wedding. The bridegroom's party would leave the following day. The shehnai* was playing at the door, and wedding songs could be heard from indoors. In the verandah there was a long row of string beds on which the wedding guests were snoring.

Madan Singh enquired, 'The vehicles that you despatched, will they reach Amola by tomorrow evening?'

Padam Singh: No, they should get there earlier. Amola is close to Bindhya Chal. I despatched the vehicles in the forenoon.

Madan Singh: Then what do we need to take from here?

Padam Singh: I don't think you need to carry anything for the function. But perhaps some food should be taken, because it will be troublesome if the meals are served late there.

Madan Singh: How much was settled for the dance? There are two dancing girls, aren't there?

Padam Singh turned pale. He had been fearing this question. He bent his head humbly, and said in a low voice, 'I have not arranged for the dance.'

Madan Singh was startled, as though he had been pinched hard. He said, 'You've ruined it all! What were those arrangements you were making from Jenva, then? Couldn't you find the time, or did the expense frighten you? That is why I had informed you four days earlier. Those who invite Brahmins must be able to afford a proper welcome. If you were afraid of the expense, you should have written so to me frankly, and I would have sent you the money. I am not yet dependent on anyone, thank God! Now tell me how it can be arranged? We are going to the home of a respectable man, what will he think? His friends and relatives would have come from afar. People from distant villages will arrive to join the bridal party. What will all of them say in their hearts? Oh, God!'

Munshi Beja Nath was part owner of Madan Singh's village. He looked at Madan Singh meaningfully and said, 'Not in their hearts, they will say it out aloud. They will proclaim it, they will abuse you to your face. Besides, they will slander you in all the neighbourhood areas. Has there ever been a function without the courtesans' dance? At least I have never seen one! Perhaps Bhaiya* forgot to arrange it or didn't have the time to do so.'

Padam Singh replied shamefacedly, 'No, that is not what happened.'

Madan Singh: Then what was the reason? You must have thought that the whole burden was falling on you. But I can tell you truthfully that I did not write to you with the intention of weighing you down with the expense. I don't believe in taking advantage of others.

Padam Singh could not bear his brother's reproaches any longer. With his eyes brimming with tears, he said, 'Bhaiya, for God's sake don't think of me in this way. I would willingly give my life for you. It is my heart's desire to serve you. The reason why I did this was that these days there is a drive among the educated people in the city to stop the custom of dance, and I have joined in this effort. Therefore, I lacked the courage to go against my declared principles.'

Madan Singh: I see. I am glad that people are becoming more enlightened. Personally, I think that this custom is indecent, but still, I don't want to look foolish. When other people drop the custom so will I, but why should I be the first to do so? I have just one son, I want to have a lavish wedding for him. Once he is married, I'll join your campaign. For now let me follow the old customs. When God gives you a son and it is time for him to be married you can practise the new ways and I'll make no objection. If it is not too much trouble, take the morning train and bring the two dancing girls to Amola. I am telling you to do this because you know the people there and can make the arrangements at less cost. Others would rob me.

Padam Singh bent his head and began to think. Seeing him silent, Madan Singh said sharply, 'Well, why are you silent? You don't want to go?'

Humbly, and with tears in his eyes, Padam Singh began to say, 'Bhaiya, please excuse...'

Madan Singh: No, no, I am not forcing you. You need not go if you don't want to. Munshi Beja Nath, can I trouble you to go?

Beja Nath: I don't mind.

Madan Singh: If you take the morning train you will reach Amola by the evening. I'll be very grateful to you.

Beja Nath: Don't worry, I'll go.

The three men did not speak for some time. Madan Singh was thinking that his brother was ungrateful. He had brought him up as his own child but on a small matter like this he had disobeyed him. Beja

Nath was thinking that Madan Singh's turning to him for help may have offended his brother. Padam Singh was just afraid—afraid of his brother's wrath. He did not dare to lift his head. On one hand was his brother's wrath, on the other the destruction of truth and principles— on one side a pitch dark valley, on the other an unscaleable mountain. Duty to his brother and duty to his conscience wrestled with one another. He was thinking, 'My brother brought me up. This body belongs to him. Yet I can't kill my conscience. But what is conscience? Merely a patchwork of circumstances and situations. And how did I become capable of accepting their influence? Through the blessing of education which was provided to me by my brother. What is the worth of my conscience when compared to his approval, or for that matter, of my fame, my principled stands? None whatsoever! I will not be branded "undutiful". I value his approval more than my conscience and my principles. But why not try to convince him once more? If he does not agree I will not shrink from obeying him.'

Having decided on this course of action, Padam Singh said to his brother humbly, 'You have always forgiven me my stupidities, please forgive me my boldness in asking you this: If you find the custom of dancing repulsive, why are you insisting on it?'

Madan Singh replied with exasperation, 'You talk as though you were not born in this country and have come from a foreign land. After all this is not the only such custom, there are many others which have to be followed even though they are distasteful, because otherwise we lose face. Singing abuse* at weddings is not a good custom, nor is expecting a dowry, but if one doesn't do what is expected, one invites ridicule. If we don't take the dancing girls with us, people will say it's because I am a miser. Nobody is going to look at my motives.'

Padam Singh: What if you spend the same money on a more appropriate cause? Nobody will complain of your miserliness then. You want to hire two dancing girls. That will cost you not less than three hundred rupees. Instead of spending three hundred, you could buy blankets for five hundred rupees, to distribute among the poor of Amola. At least two hundred people will pray for you and will praise you until the last thread in the blanket has rotted. If you don't like this idea, then perhaps you could have a good well constructed in the village. You will always be remembered then, and I'll bear the expense of the well.

Padam Singh's suggestion demolished his brother's excuse about losing face. While Madan Singh was thinking of an appropriate reply,

Munshi Beja Nath who in spite of his fear of annoying Padam Singh, could not forgo an opportunity of displaying his ingenuity and sagacity, put in, 'Bhaiya, there is a time for everything. Celebrations and charity are both good at the right time, and most inappropriate when the occasion does not call for them. Besides, it is not the educated people of the city that we are dealing with. If you distribute blankets among the uncouth villagers they will stare at you and laugh.'

Madan Singh who had been at a loss for an appropriate reply was pleased at Beja Nath's intervention. He looked at him with gratitude and said, 'That is true. How can anyone appreciate monsoon music in spring?* There is a time for everything. And that is why I want you to leave early tomorrow morning and settle the matter with the two dancing girls.'

Padam Singh thought, 'These people will eventually do what they want, but let's see how they justify what they do.' It hurt him that his brother seemed to trust Munshi Beja Nath more than himself. He said boldly, 'So why should we assume that a marriage can be celebrated only with a feast? On the contrary I think that it is the best possible occasion for charity and good works. Marriage is a sacred duty, a spiritual contract. It is the moment when we take on worldly responsibilities, when we are fettered with worldly relationships and the ensuing worries, when we bow our head and accept duties and restrictions. It is our duty to respect such a sacred custom and to observe it with gravity. It is heartless of us to be feasting, and celebrating with dance and music the moments when one of our kin is being burdened with his heavy, new duties. If unfortunately this has now become the custom, it doesn't follow that we must slavishly observe it. Education should make at least this much difference, that we stop giving prime importance to such superficial delights where spiritual matters are concerned.'

Beja Nath looked at the floor. Madan Singh was gazing at the sky; he had found his brother's speech convincing, but truth and principles are powerless before tradition. He was afraid that Beja Nath would not be able to come up with a suitable rejoinder. But Munshiji did not want to give up yet. He said, 'Bhaiya, you are a lawyer. I do not have the skill to argue with you. But it is true that if we stop observing traditions we will be disgraced; whether those traditions are appropriate or not. After all our elders were not ignorant barbarians. They must have had good reasons for starting those traditions.'

This argument had not occurred to Madan Singh. He was very pleased with it. He looked appreciatively at Beja Nath and said, 'There is bound to be some wisdom in the customs that were established by our elders, even if we are not able to perceive it. Some modern gentlemen who pride themselves on putting an end to those customs despise our ancestors, although whatever knowledge, culture, and dignity we possess today, was all earned by them. Some of those who have the modern mentality are against wearing the sacred thread.* Others insist on cutting off the pigtail.* Yet others want widows to remarry.* There are even those who want to get rid of the caste system. These gentlemen feel that if such distinctions are removed, India will be governed by considerations of unity and harmony. I am afraid I don't fancy such ideas. Those who want to believe in them are welcome to do so. For my part, I like the old ways. And if I live long enough I would like to see how well these European seedlings take to our soil.

'Our ancestors considered agriculture the best profession. But people nowadays run after mills and machines simply because they want to ape Europe. But the day will come when the Europeans themselves will pull down their mills to dig fields in their place. What after all is the importance of a mill-hand when compared to a free tiller of the land? What kind of a country is it where people starve if food is not imported? Countries that are governed by such ideas cannot be our models. The value of industry will last as long as there are weak, insecure nations in the world. The Europeans are rich because they sell their cheap industrial products to these countries. They sell their goods at the point of the bayonet. But when these nations awaken, Europe will lose its eminence in the world. Its wealth will disappear. We do not say, "Do not learn anything from Europe". No, they are the masters of the world today and they have many excellent qualities. Learn from them their good qualities, but don't learn the bad ones. Learn from them diligence, but not formality; learn their determination, but not their haughtiness. Do not add to the list of worldly goods that you need, in imitation of the West. Do not become a slave to your own desires. Do not believe in selfishness. Do not crush the poor. Our rites and customs suit our circumstances and our needs. They need no foreign patches.'

Madan Singh said all this with the authority of a learned man divulging his profoundest thoughts, although their substance was no more than hearsay, most of which he himself did not understand.

Padam Singh listened to him with philosophical patience and the fear that the discussion might turn into a dispute, signs of which he thought he could see. So, he thought it best to close the subject, and said very gently, 'You are right in saying that we should not follow Europe blindly, but pardon me for saying this: that circumstances have changed a great deal since our culture had its beginnings. These changes were bound to affect our way of life. But I don't want to prolong this discussion. I will take the train tomorrow morning and bring the dancing girls if you want me to do so. Why trouble Munshiji? If he goes, his work will suffer. Come with me Munshiji, let us go out. I need to talk to you.'

Madan Singh: So why not talk here? I'll go away if you like.

Padam Singh: No, don't do that. There are just a few questions I need to ask Munshiji. 'Tell me, how many people will come to watch the proceedings in Amola? About a thousand?'

Beja Nath replied cautiously, 'Probably somewhat more than that.'

Padam Singh: How many of them do you think would be poor farmers, and how many affluent landowners?

Beja Nath: Mostly poor farmers, but the landowners would not be fewer than two or three hundred.

Padam Singh: Do you agree that a man wants to satisfy his physical needs before seeking entertainment?

Beja Nath: Yes, I agree with that. How can a starving man ask for recreation?

Padam Singh: Then you will have to agree that if these farmers are given blankets or clothes it would make them happier than if they are shown a dance.

Beja Nath: No, I don't agree with that. Most farmers will not accept even a hundred blankets. They would rather attend the function, and if they find it insipid they will return home disappointed.

The series of Socratic questions that Padam Singh had devised, was interrupted. He perceived that Munshiji had got the drift of his argument, and so he decided to change his tactics. He said, 'All right, forget that. Now tell me, do you agree that the goods you see in the market and their quantity, are determined by the demand ?

Beja Nath: Yes, certainly.

Padam Singh: So customers are responsible for bringing the goods they prefer into the market. If nobody ate meat, a goat would not be slaughtered?

Beja Nath was conscious that Padam Singh was laying another trap for him but so far he had not been able to fathom where the catch lay. He replied cautiously, 'Yes, that's true.'

Padam Singh: If you believe that, you must agree that people who hire dancing girls for their functions, paying them generously so that they can afford a lavish lifestyle, are no less sinning than the butcher who slaughters a goat. And that if I had not seen lawyers ride grandly in carriages I may never have become a lawyer.

Beja Nath replied, laughing, 'Bhaiya, you manage to get your own way, no matter how. But what you are saying is right.'

Padam Singh: So then it is not difficult to understand that we are responsible for the fate of all these women of the streets, who have sold their virtue. We are responsible for all the depravity and baseness which exists as a result of their shamelessness. Those thousands of families who have been destroyed by this immorality will hold us accountable before God. Therefore we should have no hesitation in discarding a custom which can lead to such dangerous consequences. Thought is the precursor of intentions and action. Who can say that such functions have no influence on our thoughts? Isn't it wise to avoid actions which are bound to result in ruin?

Madan Singh was listening carefully to his brother's speech. His agreement with its contents could be read from his face. He had not had the advantage of higher education which makes a man a nonconformist and a freethinker. No, he was a man of ordinary intelligence and perceptions, and no quibbler. He smiled at Beja Nath and said, 'Well, Lala Beja Nath, what do you say now? Can you think of an escape route?'

Beja Nath: No, I can see no way.

Madan Singh: At least put up a substantial defence!

Beja Nath: I might have, if I had read law. Bhaiya Padam Singh, what reply would you have made if you were me?

Padam Singh: (laughing) There is no dearth of replies. Any student of philosophy can prove that black is white, the sky is earth, and light is darkness. All this is child's play for a philosopher.

Madan Singh: I concede that such functions do have a bad influence on thought. In my youth whenever I returned from a function of that kind, I used to talk of the prostitutes, their looks and gestures, and the singing and dancing, for months afterward. I couldn't get it all out of my head.

Beja Nath: Then Bhaiya, do as Padam Singh says. At worst, a few people will laugh at you; it does not matter. But do have the well constructed.

Padam Singh: I'll have the foundations dug straight away.

– 4 –

It was a day in the monsoon season. Black clouds covered the sky. Pandit Oma Nath was standing on the bank of the Ganges near Chanar Gadh, waiting for the boat. He had just returned from a tour of several districts, and at this moment was preparing to go to a village near Chanar because he had heard that there was a young man there who could make a suitable bridegroom for Shanta. The boat was still anchored on the other side of the river. Oma Nath was angry at the boatmen. But more than the boatmen, Oma Nath was infuriated by the passengers who took their time coming to the boat.

When it grew late, Oma Nath called out to the boatmen on the other side. But his voice was in no mood to reach the boatmen. It played with the ripples and was drawn into them.

Suddenly Oma Nath saw a sadhu come towards him. He was a tall man, wide-chested, and red-eyed. His hair was matted after the fashion of holy men, and he was wearing a row of large beads which hung on his chest. In one hand he was holding a hookah bowl and in the other a pair of iron tongs. A deer skin* was wrapped around his back. He came and stood by the river; he too wanted to cross it.

It occurred to Oma Nath that he had seen the man before, but he could not place him. A curtain seemed to have fallen in his mind, cutting off his memory.

The sadhu who was gazing at Oma Nath, greeted him. He said, 'Maharaj, is everybody well at home? What brings you here?'

The curtain dropped, and Oma Nath recognized the man. We can change our looks but we cannot change our voice. It was Gujadhar Panday.

Oma Nath had not visited Saman since she was married. Having pushed her into an inappropriate marriage, he did not have the courage to face her. He was surprised to see Gujadhar in this garb and thought that perhaps he was deceived. 'What is your name?' he enquired.

Sadhu: I used to be Gujadhar Panday. Now I am Gujanand.

Oma Nath: No wonder I didn't recognize you, though I thought I had seen you somewhere. Why are you in this attire, and where is the family?

Sadhu: I am free of that encumbrance.

Oma Nath: Where is Saman?

Gujanand: She sits in the red-light district.

Oma Nath looked at Gujanand with astonishment, then he lowered his head in shame. After a while he said, 'How did it happen?'

Gujanand: Just as it usually happens in this world. My ill humour and cruelty, and Saman's boldness and ambition joined forces to destroy us. Now when I think about it I realize that I made a grave mistake by marrying a girl from a higher social class.

My mistake was that being poor I could not pamper her, that I should have made up for this hardship to her with love and gentleness, but I did not. Instead, I kept her deprived of food and clothing and expected hard work from her to which she was not used to. And she could never get used to it, but I made her work and rebuked her for the smallest lapses. I know now that it is I who is responsible for her ruin. Between beauty and elegance there is the same bond as there is between a flower and its fragrance. She did not, and could not, love me, but she looked after my needs. A pauper is ecstatic if he finds a treasure, similarly, a man who finds a beautiful wife is intensely jealous and suspicious. I was no different. I always distrusted her and made her suffer for my mistrust. Maharaj, when I think of my cruelty to her, I want to drink poison. I am atoning for those brutalities. For two or three days after she left, my state of mind did not change. I was drunk on the power I had wielded. But after the intoxication subsided, I began to find it hell to live in the house. I could not bring myself to step inside it. So I became a worshipper at a temple. This at least saved me the trouble of cooking my own food.

Some sadhus used to come to the temple regularly and I had the opportunity to spend time with them. Listening to their wise and spiritual conversation opened my eyes. I then decided to adopt this garb. Now I wander from village to village, and help the poor as best I can.

Are you coming from Benaras?

Oma Nath: No, I am coming from a village. Saman has a younger sister. I am looking for a husband for her.

Gujanand: This time you should find a better match.

Oma Nath: There is no shortage of suitable matches, but one has to be able to afford them.

Gujanand: How much money would you require?

Oma Nath: One thousand for the trousseau alone, and more for the wedding expenses.

Gujanand: Please settle the marriage. I'll find one thousand rupees. As a sadhu it is not difficult to do people out of their money. I'll see you in Amola in the next three or four days.

The boat arrived and the two men got in. Gujanand started talking to the boatmen but Oma Nath was brooding. He was telling himself, 'I am Saman's murderer.'

– 5 –

Pandit Oma Nath had arranged Shanta's marriage with Sadan Singh. He did not tell Janhvi about the money Gujanand had given him for fear that she might put it away for her daughters. Janhvi could not be subdued by admonishment, so he had no option but to go along with her. He had settled the match for one thousand rupees but was worried at the thought of finding the money for entertaining the wedding party. At least a thousand rupees would be needed, but he could think of no way to find that sum. However, he was happy to think that Shanta would be married into a good family where she would live comfortably, and Gangajali's spirit would be pleased with him.

When only three months were left for the wedding and he could not procure the money, he stopped worrying about it. He decided he would find an excuse to quarrel with the bridegroom's party. Those people would take offence and go back. But the marriage would have taken place and the girl would live in comfort. He thought he would manage it all so cleverly that the whole blame would fall on the bridegroom's party.

It was a week since Pandit Krishan Chandr had been released from prison. But Oma Nath had not been able to consult him about the wedding. He was ashamed to face his brother-in-law.

There was a marked change in Krishan Chandr's character. He had lost his dignity and elegance. Physically he was thin and weak, though he had acquired a certain nimbleness of limb. Frequently, at night, he could be heard sighing and moaning. And in the middle of the night he would turn and toss in his bed, singing doleful songs. In his eyes there was now a lewdness and roguishness. He inspired such dread in Janhvi that she avoided him.

It was the winter season. The village women were out working in the fields. Krishan Chandr would stroll up to the fields, and tease the women. As the brother-in-law of someone who was part of the village community, he was entitled to banter with the women,* but his remarks were so shameless and his glances so meaningful, that the women would hide their faces in shame and later they would complain to Janhvi. The truth was that Krishan Chandr was beset with rebelliousness.

There were many educated and respectable people in Amola, but Krishan Chandr avoided their company. Instead, he could be seen every evening smoking pot with the riff raff, describing to them his experiences of prison life. His conversation was studded with coarse and obscene words.

Oma Nath was much respected in his village. His brother-in-law's indecent behaviour mortified him and he prayed to be delivered from him. Even Shanta began to feel uncomfortable and fearful in her father's presence. When the village women described Krishan Chandr's bold behaviour to Janhvi, she was overwhelmed with mortification and would immediately leave the room. She could not understand what had come over her father. How had a man who was courteous and dignified, suddenly undergone such a sea change?

A month passed in this way. Oma Nath was annoyed that Krishan Chandr showed no concern for Shanta's forthcoming marriage. After all she was his daughter, not Oma Nath's. The least he could do was to try and earn some money, instead of which he was wasting his own life and spoiling Oma Nath's.

One day Oma Nath threatened Krishan Chandr's companions. He said, 'If I see you smoking pot with him again, I will sort out every one of you!'

As Oma Nath was held in high esteem in his village, Krishan Chandr's cronies were unnerved by his wrath. So when Krishan Chandr visited them the next day they told him, 'Maharaj, please don't come here again. It will cost us dear to be on the wrong side of Pandit Oma Nath.'

Krishan Chandr went to Oma Nath in a furious temper. 'I think that you don't want me to stay in your house', he said.

Oma Nath: It's your house too. Stay as long as you like. But I don't want you to disgrace yourself and me by keeping company with disreputable people.

Krishan Chandr: So what company should I keep? None of those who are considered respectable here want to be seen with me. All of them think I am of no consequence. I can't bear such treatment. But can you name one man from among them who is perfect? They are all treacherous, corrupt, and enemies of the poor. I don't consider myself worse than they. I am paying for my lapse. They are not paying for anything. The difference between them and me is that they commit a thousand sins to hide one sin. In this respect they are worse than I am. I don't want to be demeaned before such hypocrites. I want to be with those who in spite of my present condition, accord me proper respect; who don't consider themselves my superiors, and who are not like the proverbial crow that pretended to be a swan. And if my attitude causes you embarrassment, I don't want to force myself on you by continuing to live in your house.

Oma Nath: As God is my witness, this was not the reason why I rebuked those people. As you know I need to keep in touch with government servants and high officials. Your unrestrained behaviour puts me in a difficult position in their presence.

Krishan Chandr: So tell your high officials that at his worst Krishan Chandr is better than they are. I was a government servant once and I know their type well. They are all cunning thieves, robbers. They stand in sin up to their necks. Such a base lot cannot teach me good behaviour!

Oma Nath: You may not care for the officials, but my livelihood depends on their goodwill. How can I oppose them? You yourself used to be a police inspector, don't you know that the Police Inspector here, in whose charge you were, would write an unfavourable report about you if he sees you in such disreputable company? That would not only ruin you, it would drag me down too.

Krishan Chandr: Who is the Police Inspector here?

Oma Nath: Sayed Masood Alam.

Krishan Chandr: That traitor! Utter fraud! Arrant rascal! As a head constable he used to be my subordinate. I saved him from going to prison once. Let him come here, I'll sort him out!

Oma Nath: If you want to start a row, please leave me out of it. It won't hurt you but it will crush me.

Krishan Chandr: That's because you are a respectable man while I am a vagabond. Don't make me say this, but you are so proud of your respectability even though you pimp for police inspectors!

Oma Nath: I may be base, sly, treacherous, and a pimp; but seeing how useful I have been to you, you might curb your tongue.

Krishan Chandr: Your good turns! You destroyed my family! Aren't you ashamed of boasting about what you did for me? You killed my wife, you married one of my daughters to some pauper, and you are making my other daughter work for you like a slave! You call that doing me a favour? On the pretext of conducting legal activity, you defrauded my poor innocent wife of all her money and then brought her to your home to ruin her. And now you boast of your kindness to me!

A self-centred person finds ingratitude more disheartening than anything else. He may not be expecting expressions of gratitude, but just the thought of his beneficence gladdens his heart. Even if he disapproved of professions of gratitude, he would expect to be the object of silent appreciation. Oma Nath was thinking, 'The world is so mistrustful. How hard I worked for his release, I used to haunt the courts, flatter the lawyers, cajole the officials. I spent so much of my own money too. And this is what I get for it! I looked after three women for all these years, spent months looking for a match for Saman, and have been working hard for many months to get Shanta married. The soles of my feet are blistered from all the walking I have done. I have lost weight and lost my appetite in my anxiety to find the money for the marriage. And this is how all my toil and trouble are rewarded! Truly, the world is blind. Even good deeds can dishonour one here.' These thoughts affected him so much that his eyes filled with tears. He said to Krishan Chandr, 'Brother, everything I did was done with the best intentions. But if it is God's will that all my effforts come to naught, then so be it. Perhaps I have robbed you, swindled you of all your wealth. So punish me as you like, what else can I say?'

Oma Nath would have liked to say, 'Enough is enough. Get off my back now. Manage Shanta's wedding yourself.' But he was afraid that in his fury Krishan Chandr might really take the girl away, and so he decided to pocket his own feelings of hurt. In the good-hearted weak, anger turns into compassion. What can a kind man do except keep quiet when abused by a beggar? In fact he feels sorry for the wretched man.

Oma Nath's fortitude had the effect of cooling Krishan Chandr's wrath. But the two men did not speak much. They sat quietly, lost in their own thoughts, like two dogs facing each other in silence, after a fight. Oma Nath was thinking, 'I am glad that I did not retaliate. If I had, the world would have blamed me.'

Krishan Chandr was thinking, 'I wish I had not brought up the past.'

An unnecessary bout of temper usually awakens one's conscience. Krishan Chandr began to see where his duty lay. He asked Oma Nath, 'You have arranged Shanta's marriage, haven't you?'

Oma Nath: Yes, with Pandit Madan Singh's son, in Chanar.

Krishan Chandr: From the sound of his name he seems to be a respectable person. How much is the dowry?

Oma Nath: One thousand rupees.

Krishan Chandr: The rest of the expenses too will probably amount to that much?

Oma Nath: Naturally.

Plucking up his courage Krishan Chandr enquired, 'How will you arrange for so much money?'

Oma Nath: God will find a way. I already have one thousand; another thousand is needed.

Krishan Chandr articulated with remorseful humility, 'You see the condition I am in...' falling tears interrupted his speech.

Oma Nath replied in a consolatory tone, 'Don't worry, I'll manage everything.'

Krishan Chandr: May God reward you for your generosity. I was not in my senses just now, Brother. God knows what I said to you. Please do not take offence. Life in that hell has killed my spirit. I have no feelings left. Under that evil spell an angel could turn into an ogre. I do not have the strength to take on the heavy burden of a marriage. You have saved me from drowning, but it is not seemly that I should let you carry the whole burden and remain idle myself. Permit me to go where I can earn some money. I will go to Benaras tomorrow. I have many acquaintances there but I'd rather not stay with them. Where does Saman live in Benaras?

Oma Nath's face turned pale. He said, 'Stay here till the wedding. After that, go where you will.'

Krishan Chandr: No, let me go tomorrow. I'll return a week before the wedding. I'll stay at Saman's for a few days and look for employment. Where does she live?

Oma Nath: I can't remember very well. I have not been there for a long time. In any case people in the cities keep changing houses. I don't know where she is now.

At dinner that night, Krishan Chandr asked Shanta for Saman's address. Not understanding Oma Nath's warning signals, Shanta gave the full address.

– 6 –

There were eighteen members of the town's municipal board. Eight of them were Muslim and the rest were Hindus. The membership consisted predominantly of educated men, so Padam Singh was sure that the motion for the expulsion of prostitutes would be approved. He had met all the members individually and resolved their doubts and objections. But there were some from whom opposition was expected. All these people were respectable and influential bankers and traders. For this reason Padam Singh feared that a section of the educated members might give in to their pressure. Among the Hindus, Seth Balbahadar Das was the leader of the opposing party. Among the Muslims it was Haji Hashim. As long as it was Bithal Das who was running the campaign, nobody had paid much attention to it. But ever since Padam Singh and some others had joined it, Sethji and Haji Saheb had become very apprehensive. They knew that the proposal would soon be put before the Board, so both these gentlemen were taking measures to resist the attack. First Haji Saheb collected the Muslim members. Haji Saheb was very influential. He was considered the leader of the Muslims in the city. Of the other seven Muslim members, Maulana Tegh Ali was a trustee of a (shia) Imambargah;* Munshi Abul Wafa was the manager of a perfume and oil factory which had many shops in the larger cities; Munshi Abdul Latif was a big landlord, who spent most of his time in the city; was a patron of poets and a poet himself. Shakir Beg and Sharif Hasan were lawyers with superior ideas on culture; Syed Shafqat Ali was a pensioner and an ex Deputy Collector; and, Khan Sahib Shuhrat Khan was a prominent doctor. The last two were retiring people but they were not narrowminded. They were both people of firm principles and were well-respected by their compatriots.

Haji Hashim was speaking. He said, 'My dear countrymen, have you observed their latest ploy? They are past masters at dissimulation.

I have come to mistrust them so much that if salvation lay in believing them I'd forgo it.'

Munshi Abul Wafa said, 'By the grace of God we have become aware of our interests. This is an attempt to reduce our collective strength. Ninety per cent of the prostitutes are Muslim. They fast, they mourn for the Imam, they observe the Prophet's (PBUH) birthday, and the saints' anniversaries. We have nothing to do with their personal lives. It is up to God to punish or reward them. We are only concerned with their numbers.'

Tegh Ali: But is their number so large that it can affect our collective vote?

Abul Wafa: It is bound to make some difference. Look at them, they are ready to admit the Domes, who are low caste entertainers, to their side, when we know very well that they consider them despicable, and baser than animals. But for the sake of their political interests they have accepted them. Domes are numbered among professional criminals. Theft, highway robbery, murder are their common pursuits. But if there is any attempt to exclude them from the Hindu party, a hue and cry is raised. The holy Scriptures are quoted to prove that they belong to the Hindu religion. We should learn these tactics from them.

Syed Shafqat Ali said gravely, 'The government has allotted separate areas to these criminal castes. The police keeps an eye on them. I myself used to write reports on their activities when I was in service. But I don't think that any responsible Hindu ever objected to the government's policy on them, although in my view the crimes of theft and murder are not as repulsive as that of prostitution. Even a Dome woman is expelled from her community if she becomes a whore. If a Dome accumulates enough wealth he can buy his pleasure from the beauty mart. God forbid that we are reduced to this, just for the sake of serving political interests.

As for the religious observances of those prostitutes, if on the basis of it God decided to send the whole Muslim nation to paradise, I would prefer to go to hell. If on the basis of their numbers we were offered the sovereignty of this country I would not accept it. My personal view is that not only should they be expelled from the centre of the city, they should be turned out of the city limits.'

Hakim Shuhrat Khan: Sir, if it were within my power I'd expel them from India. I'd settle them in a separate island. I frequently come in contact with their clients, and if it didn't affect my religious faith,

I'd say that whores are the reigning deities where cholera and the plague are only abject subjects. Cholera finishes off its victims in two hours, plague does the same in two days, but these denizens of hell induce a slow and torturous death. Munshi Abul Wafa considers them the houris of paradise but in fact they are black cobras with poison in their eyes. They are the lakes that give rise to rivers of crime. How many good women there are who are caught in a living death because of them! How many respectable young men are weeping tears of blood on account of them! It is our misfortune that shameless prostitutes call themselves Muslim.

Sharif Hasan said, 'There is nothing wrong in their calling themselves Muslim; what is not right is that like the Hindus, Muslims too have done nothing about reclaiming them. It is true that our divines in their green turbans, with antimony in their eyes, and their hair carefully combed, sit at their table to eat, smoke tobacco, and chew fragrant paan, all for the cause of restoring them to religious favour. But Islam's reformatory prowess ends there. It is in human nature to regret one's misdeeds. These misguided women are bound to grieve over their condition, once their intoxication is over. But at that point their repentance can serve no purpose as no other means of earning a livelihood is left open to them. If these girls could be legally married and assured of financial support, I am certain that at least seventy-five per cent of them would gladly opt for a respectable life. No matter how vile we ourselves are, we want our children to lead blameless lives. Turning whores out of the city will not bring about reform. From this point of view I can dare to object to their expulsion, though I cannot object to their eviction on the basis of political gains, because I cannot approve of an action on the grounds of collective good unless I am convinced that it is morally sound as well.

Tegh Ali: Watch out sir, they might brand you an infidel! This is the age of political interests, so don't talk of truth and justice. If you are a teacher, fail your Hindu students; if you are a revenue official, tax the Hindus unjustly; if you are a magistrate, give harsh sentences to the Hindus; if you are a police inspector, involve them in false lawsuits; if you are an investigator, falsify the statements of the Hindus; if you are a thief rob a Hindu household; if you are amorous, ravish a Hindu beauty. Only then will you be known as a benefactor of your people and a captain of the nation's boat.

Haji Hashim was perturbed, and Munshi Abul Wafa scowled. Tegh Ali, a name that translated as 'the sword of Ali' had demonstrated the

sharpness of his sword. They were about to respond when Shakir Beg
spoke up. He said, 'Brother, this is not an occasion for sarcasm. It is a
mutual consultation, not a forum for polemics. A sharp tongue is
poison for harmonious discussions. In my opinion courtesans are
neither superfluous nor evil to the interests of civilization. When you
build a house you must build sewers; if you don't, the walls of your
house will collapse in a few years. Prostitutes are the sewers of society;
and just as sewers are built in an inconspicuous part of the house, so
this community should be removed to a far corner, away from the
environment of the city.'

Munshi Abul Wafa who had cheered up when he heard the first
part of this speech, became dejected when he heard the metaphor of
sewers. Haji Hashim looked despairingly at Abdul Latif and said, 'Do
you have something to say, or are you too going to retreat in the name
of national unity?'

Abdul Latif replied, 'What have prostitutes to do with unity or
disunity? I am a votary of peace and harmony. At this moment I
cannot decide whether I am awake, or asleep and dreaming; because I
have been listening to these learned people hold forth in support of a
proposal that is so inane. How can I be awake? You have no objection
to shops that sell soap, leather, or kerosene. Fabric, crockery, and drug
shops are all there in the market; you are far from considering them an
encumbrance. Yet, do you believe that Beauty is so worthless that it
should be banished to a dark corner of the earth? Would a garden
deserve to be called a garden, where the rows of pine are in one
corner, the beds of roses and jasmine in another corner; where there
are neem and acacia trees on both sides of the avenues, in the centre a
pipal tree, and a grove of cactus by the pond; where crows and kites
shriek from the trees, and the nightingales sing dirges from some dark
corner? I do not consider this proposal fit to be made the subject of a
serious discussion.'

Haji Hashim smiled. Abul Wafa's eyes were shining with delight.
Others listened to this facetious speech with a philosophical smile.
But Maulana Tegh Ali was not so patient. He said sharply, 'Well sir,
why not propose to the Board that the municipality should arrange a
fair, right in the middle of the marketplace. Those who go to the fair
should be awarded a certificate of pleasant temperament by the
government. I think that there will be many who approve of this
proposal, and it will bring great fame to the proposer. After his demise,
his death anniversary will be celebrated yearly at his tomb, and lying

in it, he will be entertained to Beauty, and the charming songs of the beauteous.'

Munshi Abul Wafa's face went red. When Haji Hashim sensed that unpleasantness might be brewing, he said, 'I used to hear that there is such a thing as principles, but I've discovered today that it is merely a notion. It is not very long since you gentlemen led a deputation for Islamic scholarships. You have also given suggestions on the religious instruction of Muslim prisoners. If I am not mistaken you gentlemen are always in the forefront of such schemes; but today I notice that there is this sudden revolution in your thinking. Well, you are entitled to your capriciousness. But as for me, I am not that gullible. And I have made it a rule of my life to oppose every suggestion of my countrymen because I have no expectation of altruism from them.'

Abul Wafa stated: And I'll add that I can believe the sun was seen at night, but I can never be convinced of the good intentions of Joseph's brothers.

Syed Shafqat Ali: Haji Saheb, sir, you consider us unprincipled and devious—that is not a mature judgment. Our principles are the same as they have always been and will always be. They are: to establish the dignity of Islam, and to work for the good of our co-religionists by every lawful means. If there is a scheme by which my co-religionists benefit but the rest of our countrymen suffer, I'd go ahead with it, but a proposal which benefits all communities equally will not be opposed by us, because we don't believe in opposition for the sake of opposition.

It was late, and the gathering dispersed. The discussion had served no useful purpose: none of the participants had changed their minds. Haji Hashim began to have some doubts about his success, of which he had been very sure.

– 7 –

When the Hindu members who had opposed the scheme found out about the conclusions of the Muslims' meeting, they were disappointed. There were just ten Hindu members, whose Chairman was Balbahadar Das. Dr Shyama Charan was the Vice Chairman, and Lala Chaman Lal and Dina Nath Tivari were representatives of the business community. Padam Singh and Mr Rustum Bhai were lawyers. Romesh Dutt was a college professor. Lala Bhagat Ram was a contractor, Pandit Parbhakar Rao was the Editor of the Hindi newspaper, *Jagat*, and Kanwar Anarudh Bahadur Singh was the biggest landlord in the district. Most of the shops in the marketplace belonged to Balbahadar Das and Chaman Lal, while in the rice market many houses were owned by Dina Nath. These three gentlemen were opposed to the suggestion, and as Lala Bhagat Ram's business depended on Chaman Lal's patronage, he too was on their side. Parbhakar Rao, Romesh Dutt, Rustum Bhai, and Padam Singh supported the proposal. Dr Shyama Charan and Kanwar Sahib's views were still not known and both sides were hoping to enlist their support as it would tip the scales in their favour.

Pandit Padam Singh had still not returned from the wedding, and taking advantage of his absence, Balbahadar Das invited all the Hindu members to his mansion. The purpose of this move was to win the doctor and Kanwar Anarudh Singh to his side. Parbhakar Rao was a known adversary of the Muslims, so he hoped to secure him also by giving the whole question a communal slant.

Dina Nath Tivari said, 'Our Muslim brothers have shown much boldness in this matter. I did not expect it of them. But as you probably know, the secret of it is that they have adopted the policy of killing two birds with one stone. On one hand they acquire a reputation for reforming society, on the other they ensure that losses are inflicted on the Hindus. How can they let such an opportunity go?'

Seth Chaman Lal: I have nothing to do with politics, but I must say that in this matter our Muslim brothers have clamped their vice around our necks. Most of the houses in the red-light district and in the marketplace belong to Hindus. If the Board adopts the proposal, we Hindus bear the losses and they get the kudos! They are snakes in the grass. Only recently they attacked Hindus on the pretext of usury. Now they have come up with this new excuse. It is a pity that some of our own brother Hindus, in their greed for cheap popularity, have become puppets in the hands of nationalism. They do not realize how much their deviation will cost the Hindu people.

The question of usury had been discussed in the local council, and Parbhakar Rao had opposed it vigorously. By bringing it up in the context of the prostitutes' eviction Chaman Lal had hoped to sway him. Parbhakar Rao looked helplessly at Rustum Bhai, as if to say, 'Save me from my friends.' Rustum Bhai was an outspoken man. He stood up and confronted Chaman Lal. He said, 'I am very sorry to see that some people are turning a cultural issue into a confrontation between Hindus and Muslims. The issue of usury was treated in the same way. To transform a matter of national concern into a bone of contention may have served the interests of some moneylenders, but it is difficult to measure the harm it would cause to national unity. It is true that if the proposal is accepted, Hindu moneylenders would be the worst hit, but Muslims too would have to suffer on account of it. In the marketplace and in the red-light district, Muslims own quite a lot of houses. We should not doubt the motives of our Muslim brothers out of prejudice and for the sake of opposing them. Their decision was taken in the spirit of public welfare. If Hindus have more to lose as a result of it, that is another matter. I am sure that if Muslims had more houses in that locality their decision would still have been the same. There must be hardly anyone in this gathering who can claim to be ignorant of the moral and social evils, to correct which this proposal has been put forward. If you are aware and sincerely concerned about these ills, you should have no hesitation in supporting this move. The subject of morality could be a matter of life and death; mere financial loss can have no significance here!'

Parbhakar Rao was relieved. He said, 'This is just what I was going to say. It is a moral and cultural problem. The financial angle is certainly not its most important feature. Already Hindus have a bad name for their tight-fistedness, and if we give so much importance to considerations of money in this matter, the rest of our fellow countrymen will have another excuse to taunt us. It is as clear as daylight that the courtesans' quarter is a shameful element of our society and therefore it should not be allowed to exist in the prominent areas of the city.'

Kanwar Anarudh Bahadur Singh looked at Parbhakar Rao and said, 'Sir, you are preoccupied with the affairs of your newspaper, so you have no time to enjoy life. But carefree people like me need some form of entertainment. Evenings are spent in playing polo, the siesta takes care of the afternoon, mornings are devoted to meeting officials, but how can we while away the time between the evening and ten o'clock at night? Today you suggest that courtesans should be removed to some remote place, tomorrow you will be proposing that there should be no dance performances by them within the municipality limits, without prior official permission. If that happens it will become impossible for us to live in the city!'

Parbhakar Rao replied, smiling, 'Are there no pastimes except polo, and dance and song? You could read something.'

Kanwar Sahib: We consider reading beneath our dignity. We don't need to become bookworms. We have already been taught everything that is needed to lead a successful life. I am skilled at French and Spanish dancing. You must have heard of them or at least read about them in English novels. Give me a piano and I can play it so well that Mozart and Beethoven would be put to shame. I am well-versed in British etiquette and customs. I know when to don a solar hat and when to wear a turban. I read books—you will find many bookshelves, full of books, in my room. But I don't hide behind those books. If this proposal of yours is implemented, I and others like me will be devastated. It is my belief that unless a man has served his apprenticeship in the school of Beauty he cannot acquire elegance. In the olden days people learned suavity, eloquence, and charm in this same school,* and the reason why today's educated people are so dry and unrefined is that they have not had the advantage of being moulded there. Even a great master like Tansen* leaves them cold. Your idea can only strengthen such philistinism. I can never subscribe to it.

Kanwar Sahib's speech had the effect of resolving doubts as to which way he would vote. Dr Shyama Charan then looked at Kanwar Sahib and said, 'I intend to ask some questions in the council, on this issue. Until the government answers those questions, I will not be able to make my views known.' Saying this he read out his list of questions.

'The Government will probably not be able to answer these questions', Romesh Dutt commented.

Dr Charan: Even if they do not respond, at least the questions will have been put. What more can we do?

Seth Balbahadar Das's hopes for a victory rose. In a vigorously argued speech he dealt with the proposal from every angle. He said, 'I am not a believer in cultural revolutions. In my view society amends itself whenever it needs to. It has no need for a reform to be imposed upon it from outside. And unless the need for the amendment is universally felt by its members, no outside movement can provide it. The restrictions on going abroad, the separation between castes or races, the meaningless restrictions on food, all must go, meekly, with their heads bent, when the question of livelihood arises. I want society to be absolutely free in such matters. We are ready to give our lives for freedom. Can we achieve freedom for the country without cultural independence? When the whole nation says with one voice, "We don't want to see those faces in the parlours of the courtesans' quarter," no power on earth will be able to ignore it.'

Sethji then continued his eloquent speech in these words:
'You are rightly proud of our music. Even those who have been charmed by the music of Italy and France, are impressed by the sophistication and spirituality of our music. But it is the very community on whose displacement some of our friends here are expending their energies, who are the guardians of this heavenly blessing. Do you want to lose this precious heritage by destroying that community? Don't you know that whatever feeling we have left for religion and nationhood, whatever love for our ancestors, we owe it all to our music. If it were not for this great art, nobody would remember Ram and Krishan, Shiva and Shankar.* Even our worst enemy could not have thought of a better way of erasing all feelings of nationhood from our hearts! I don't say that this community has not supported the subversion of morals, it would not be factual to say so, but the cure for a disease lies in its treatment, and not in death. No base custom can be abolished by disparaging or disgracing it. It can be reformed through education, civility and tact.

'There is no easy approach to paradise; everybody has to pass through the eye of the needle. Those who think that on the strength of a saint's intervention they can leap into heaven are as ridiculous as those who believe that as soon as the prostitutes are banished from the marketplace, India's dark days will be over, and a new sun will begin to illuminate its skies.

'Some of our friends have looked at only the financial aspect of the matter. But I agree with Mr Rustum Bhai and Pandit Parbhakar Rao that money has no significance where morality is at stake. However, I am not prepared to bear huge monetary losses for the sake of doubtful and suspect moral gains. If you try to have this proposal passed by the Board without respecting the feelings of the affluent, then you should be prepared to face the opposition that they will undoubtedly put up in order to safeguard their own interests. I also beg to remind my affluent friends that wealth entails certain responsibilities. National movements are run on the generosity of the rich. I hope that in the course of opposing this move they will not lose all sense of balance.'

– 8 –

Just like a lazy man who wakes up when he is called but goes back to sleep when he finds nobody around, Pandit Krishan Chandr became mindless of his duty as soon as the wave of remorse had

abated. He thought, 'I am not much of a burden on Oma Nath. All he
has to provide is a little flour for my bread.' But at the same time he
had stopped smoking pot with his disreputable companions. Instead,
now, he would sit on the verandah staring at passing women. He
would agree with everything Oma Nath said, and would eat whatever
was set before him without ever asking for more. He had lost his self-
respect.

Whenever Oma Nath mentioned Shanta's marriage, he would say
with alacrity, 'Do what you like. It is all in your charge.' He would
tell himself, 'Since he is shouldering the expense he should do it his
own way.'

But Oma Nath had not forgotten his biting allegations. If you put
butter on a blister the pain will subside for a moment, but it will come
up again, soon enough, and so Oma Nath forgot Krishan Chandr's
humble apology which had appeased him momentarily, but his
ungrateful words continued to ring in his ears.

When he went to bed Janhvi asked him, 'Why was Krishan Chandr
angry with you?'

Oma Nath said complainingly, 'He said, "You have robbed me.
You killed my wife, threw one of my daughters into a dark well and
are abusing my other daughter to death".'

Janhvi: Don't you have a tongue in your mouth? You should have
said, 'I brought her because your wife had nowhere to turn.' Why,
here we do all we can to preserve the family honour and this is what
we get for it! As for them, he worked as a police superintendent for so
long but never once did Gangajali send me the smallest gift. If I had
been there when he accused you I would have set him right! He
burdened us with two young girls, the expense has beggared us. And
this is how he shows his gratitude! Why doesn't he take himself off?
What's stopping him here?

Oma Nath: He does want to go now, that's why he is asking for
Saman's address.

Janhvi: So now he wants to foist himself on his daughter, the
shameless man!

Oma Nath: No, I don't think so. He may stay there for a few days.

Janhvi: What are you talking about? He is incapable of doing honest
work. He has lost all shame. He'll dump himself there, but mark my
words, he won't be happy there for a day.

Oma Nath had not confided the story of Saman's shameful
transformation to his wife. He knew that women are incapable of

keeping a secret and that she was no exception. However when he was pleased with Janhvi he had a great urge to tell her all about it, though he held back when he thought of the likely consequences. But on that day Krishan Chandr's bitterly painful words and his wife's sympathy disarmed him, and he told her the whole story. As soon as he had finished he became aware of his mistake, but it was too late.

Janhvi had promised her husband never to divulge the secret, but it was like a burden on her chest, and she began to lose interest in her work. She was irritated with Oma Nath for encumbering her with it. She did not hate Saman, nor did she have any sympathy for her. She was not angry; merely bursting with the urge to comment on a tragic example of moral corruption. And what a strong argument it was against the education of women!

Janhvi could not deprive herself of the pleasure of disclosure for very long. Keeping the secret was a sort of disloyalty to those women who kept her informed of the smallest details of their lives. Moreover Janhvi was also very eager to hear their comments on this affair. She restrained herself for a long time, until one day Subhagi, the wife of Kabeer Pandit, came to her and said, 'Will you come with me to bathe in the Ganges?'

Janhvi was on very friendly terms with Subhagi. She told her, 'I would come with you but the angel of death is sitting at my door. For fear of him I cannot go anywhere.'

Subhagi: I am ashamed to tell you my experiences of him. If my husband hears about it he will be ready to behead him. Yesterday he was serenading my daughter with some vulgar song. Early this morning I saw him flirting with her at the well. I don't hide anything from you. You know, if something untoward happens we'll be disgraced in the whole village. He is getting old; what business does he have to behave in this way? And then my daughter is no more than one or two years older than Saman. Besides, had she been his sister-in-law it would still be forgivable, but her relationship to him is more like a daughter's, doesn't he realize this? If my Panditji hears of it there will be bloodshed. Please talk to him quietly and try to make him understand.

Janhvi could no longer restrain herself. She told Subhagi the whole story about Saman, adding colourful details of her own. When somebody tells us their secret, it is hard not to tell them ours.

The next day Kabeer Pandit's wife sent her daughter back to her husband. She had resolved in her heart to take her revenge on Krishan Chandr.

– 9 –

It was Sadan Singh's wedding day. The bridegroom's party left Chanar for Amola. There is no point in writing the details of this occurrence, as this procession was no different from other such spectacles—a strange amalgam of excessive ceremonial hospitality and painful deprivation. The palanquins were covered with brocade, but the palanquin bearers' uniforms were old and moth-eaten. Semi-naked labourers held in their hands maces and lances of inlaid metal.

Amola was twenty miles away. On the way was a river, to cross which the bridal party had to hire boats. It took hours to negotiate with the boatmen, after which, finally, they unhooked the boats. Madan Singh was very annoyed, he said to the boatmen, 'Had you been in my village I would have made you work so hard that you'd have cause to remember.' However Padam Singh was very pleased by the boatmen's boldness. He saw in it a sign of their love of independence.

In the evening, the party reached Amola. Padam Singh's people had already put up a vast and colourful marquee. There were also many tents, and gas lanterns were hung in front of them. The marquee was complete with electric lights and chandeliers, and under it were laid out a brocade rug, bolsters, and carpets. Sprays for rosewater, and khasdans* were placed in convenient places. The rumour was that several groups of dancers were about to arrive.

The proceedings began with the duvar pooja.* Oma Nath, with a bathing towel over his shoulders, was receiving the bridal party. The village women, standing under a canopy, were singing songs of welcome. The guests were scouring the bride's side for the most beautiful faces, while they were being observed with equal attention by rows of eyes on the other side. Janhvi was sad. She was thinking, 'My Chandra should have got this bridegroom.' Subhagi was dying to know which of the men was the bridegroom's father. Krishan Chandr was washing Sadan Singh's feet and thinking what a vulgar custom it was. Madan Singh was watching carefully to see how much money had been placed in the platter.

Chaos reigned as the rations began to be distributed. Some guests were asking for more, others refusing any. Someone said, 'I didn't get enough ghee', another complained, 'I have no dung cakes.' Beja Nath was demanding wine. He was angry with the whole bridal party, and several people were trying to appease him.

After the rations were distributed, people lighted the dung cakes and meals began to be cooked. The rising smoke darkened the sky and the gaslight paled.

Sadan sat on the brocade rug. The festivities had begun. Attar and paan were being passed around. The musicians picked up their instruments and the first strains of the enchanting Shayam Kalyan filled the air. Thousands of people had collected around the marquee. Some people, turbaned and clad in quilted waistcoats, with pouches of tobacco and betel-nut in their hands, were sitting on the carpets. People were asking each other, 'Where are the dancers?'

They looked into one tent and peeped into another, and said with amazement, 'What kind of bridal party is this, there are no dancers! Are they completely bankrupt? Then why bother to put up this vast marquee?' The melody of Shayam Kalyan failed to appease them. They did not care for the music, they sought the dance and the dancers. Overhearing the angry whispers, Madan Singh was infuriated with his brother. And as for Padam Singh, he was unwilling to show himself to Madan Singh out of fear and mortification.

Meanwhile people began to pelt the marquee with stones. Beja Nath ran into a tent, and some people began to abuse the miscreants. Soon there was pandemonium. People were running here and there, some were cursing, others were heading for a brawl. Suddenly, a big, powerfully built sadhu with the traditional three-pronged lance in his hand, came and stood in the marquee. His eyes were burning like two red lights and his face wore an expression of awesome authority. The gathering fell silent at once and people began to regard him with intense fascination. Who was he, and where had he come from?

The sadhu waved his weapon and spoke in a tone of reproach: Alas! there is no dance here, there is no whore. All the gentlemen are unhappy. The melody in Shayam Kaliyan is bewitching but nobody wants to listen to it. Nobody has ears for it. They all want to see the dance. Show them the dance. Or shall I show them a dance. Do you want to see the dance of the gods? Look at that tree there. See how the radiant moonbeams dance on its leaves. Look at the drops of water dancing on the lotus flower in the pond. Go into the jungle and see how the peacock dances, with its feathers spread out. Do you not like the dance of the gods? Then let me show you the dance of the demons. Look at how your neighbour, the poor farmer, is dancing when he is thrashed by the landlord. Your brothers' orphan children are dancing with hunger. In your own household, see the tears dance in the eyes of your brother's widowed wife. You don't like this dance either? Then look into yourself. Malice and its fruit are dancing there. Love and lust are swaying together. The whole world is a performance in which everybody is dancing his own dance. Have you no eyes to see this? Slaves of lust! Aren't you ashamed to mention dance? For your own good, abolish this custom, renounce sensuality, come out of this filthy slime!'

The whole gathering was silent. The people were listening mutely to the intoxicated speech of the sadhu. Suddenly he vanished, but his melodic song could be heard from behind some mango trees. Gradually the sound also faded, absorbed into the darkness, like anxiety drowning in the ocean of dreams at night.

When a group of gamblers, surprised by the police, loses its nerve, it becomes tensely active, gathering up the money, and hiding the coins. In the same way the apparition of the sadhu, his unearthly looks, and his intoxicated speech had driven a strange fear into the hearts of the people. The miscreants slipped away quietly. Those who had had regrets about coming, became attentive to the music, and some optimists ran about looking for the sadhu, but found no trace of him.

– 10 –

Pandit Madan Singh was sitting in his tent, calculating his expenses. Suddenly, Munshi Beja Nath came running and said to him, 'Brother, there is foul play here. You should never have agreed to this match.'

Madan Singh said with astonishment, 'Why, what is the matter?'

Beja Nath: I've just met a man from this village who has exposed these people. I feel devastated!

Madan Singh: Why, do they come from an inferior family?

Beja Nath: No, not an inferior family, but the matter is serious. The girl's father is an ex convict, and her sister has left her home. She is the Saman Bai who lives in the red-light district.

Madan Singh felt as though he had slipped and fallen from a tree. He said, 'Do these people have an enemy? Sometimes people say such things out of enmity.'

'Yes, it seems like one of those things', Padam Singh remarked.

Beja Nath: I don't think so. The man told me he can say it to their face.

Madan Singh: Isn't the girl Oma Nath's daughter?

Beja Nath: No, she is his niece. Do you remember that news about the trial of a police officer? He is Oma Nath's brother-in-law, now released from prison.

Madan Singh held his head in his two hands and mourned, 'Why have you done this to me, my God?'

Padam Singh: Oma Nath should be sent for.

Meanwhile Oma Nath himself arrived with his barber. He had come to fetch the clothes and jewellery which the bridegroom's family had brought for their daughter-in-law. As soon as he reached the door of the tent, Madan Singh pounced on him, and grabbing and violently shaking both his hands he said, 'You pious Brahmin! Couldn't you find somebody else to share your ignominy?'

Oma Nath looked at him with the look of a mouse caught in a cat's paws, and said, 'Maharaj, what have I done?'

Madan Singh: If I cut your throat for what you have done, I'd be sent to heaven. You were trying to thrust on us a girl whose sister has turned to vice!

Oma Nath said in a small voice, 'Everybody has his enemies. If somebody has been slandering me, please don't believe him. Call the man so he can say it to my face.

Padam Singh: Perhaps that is what it is. We should call the man.

Madan Singh gave his brother a sharp look and said, 'Even if it was an enemy who said it, just tell me whether it is true or not.'

Oma Nath: What? What do you want to know?

Padam Singh: That Saman is her real sister?

Oma Nath turned pale. His head was bent with shame. Humbly he began to say, 'Maharaj...' But Madan Singh thundered at him, 'Why don't you tell me plainly, is it true or not?'

Oma Nath again tried to speak but could not bring out anything more than the single word, 'Maharaj.'

Madan Singh now had all the proof he needed, to be convinced. He was beside himself with wrath, and sparks seemed to fly out of his eyes while his body shivered. He looked at Oma Nath with smouldering eyes and said, 'Be gone if you know what is good for you and never show yourself to me again. Traitor! Scoundrel! Hypocrite! Pretend to be a Pandit with that tilak on your forehead! I'll never drink water in your house again. Keep your girl!'

Saying this, Madan Singh went to the tent where Sadan was asleep and called out loudly to the palanquin bearers.

After he left the tent, Oma Nath turned to Padam Singh and said, 'Wakil Sahib, Please calm Panditji somehow. I will not be able to show my face to the world. You do know about Saman? That unfortunate girl has disgraced us. That was God's will. What is the point of raking up the past now? Be fair. What could I do except keep the matter secret? How could I have arranged a marriage for this girl without concealing the facts? But I tell you honestly that I myself did not know about Saman until after this match was settled.'

Padam Singh said with concern, 'If the matter had not reached my brother's ears, things wouldn't have come to this. I will try to persuade him but now that he has taken a decision it is unlikely that he will change his mind.'

Madan Singh was shouting orders at the palanquin bearers, telling them to get ready to leave. Sadan was gathering up his clothes. Madan Singh had told him everything. Just then Padam Singh said to his brother, 'Don't be in such haste, Brother. Think what you are doing. You have been deceived, it is true, but returning in this fashion will mean that you will be ridiculed as well.' Sadan looked at his uncle reproachfully and Madan Singh looked at him with incredulity. Padam Singh continued, 'Why don't you consult a few people first?'

Madan Singh: Are you suggesting that I fall into this mess, knowingly?

Padam Singh: At least you will not be ridiculed.

Madan Singh: You are young and cannot understand the gravity of the situation. Go and prepare for the departure. It is better to be ridiculed now than to have the family name blackened.

Padam Singh: But what will become of the girl, just think! It is not her fault.

Madan Singh snapped at him, 'You are a fool! Go, see to the loading of the tents. You'd be the first to taunt me about being tempted by the dowry, if something went wrong. This is not a legal problem!'

Padam Singh said in a somewhat injured tone, 'I have no hesitation in obeying you but it is a pity what will happen to the girl. Her life will be ruined.'

Madan Singh: Don't make me angry! Am I responsible for the girl?

Padam Singh remarked in a subdued tone, 'Saman Bai has nothing to do with them anymore. They have disowned her.'

Madan Singh: I told you not to make me angry. Aren't you ashamed of saying such things? You think you are a great reformer! I should marry my son to a whore's sister, should I? Are you out of your mind?

Padam Singh bowed his head in helplessness. His heart told him 'Perhaps I would have done the same under similar circumstances', but thinking of the likely consequences for the unfortunate girl he plucked up courage once more to say, 'Saman Bai has entered the Widows' Ashram.' Padam Singh was speaking with his head bent; he did not have the courage to look his brother in the eye. The moment those words left his lips Madan Singh gave him such a hard push that he fell down.

When Padam Singh raised his head he saw his brother trembling with rage. But the curses and reproaches which he was about to utter died on his lips when he saw Padam Singh fall to the ground, and in their place there was remorse and anguish. At that moment he was like a man who bites his own flesh in anger.

This was the first time in his life that Padam Singh had been thus humiliated by his brother. In all of his childhood, despite all his mischief and the many pranks he used to play, Madan Singh had never struck him, never even looked at him with anger. In his grief Padam Singh was sobbing like a child. Tears streamed down his cheeks. But in all this sorrow there was not a shred of anger. What distressed him most of all was the thought that one from whom he had had nothing but love, and never any hard words, was so agonized by what he saw as his obduracy.

It was not a blow, he was thinking, it was the manifestation of despair, broken pride, and humiliation. It was the heart's song of pain, a spark from the hidden agony, the barometer of inner suffering.

Sadan quickly lifted his uncle from the floor and looking wrathfully at his father, said to him, 'You are mad!'

Meanwhile several people came in and said to Madan Singh, 'Maharaj, what is the matter? Why have you ordered the bridal party to go back? Find a tactful compromise. Your reputations are tied together now. If they have not given you as much as you expected, be generous and overlook their blunder. God has given you everything. Their wealth cannot add much to yours.' Madan Singh made no answer.

There was confusion among the people. Men were asking one another, 'What has happened?' The crowd in front of the tent grew larger.

There were many who were from the bride's side. They were asking Oma Nath, 'Brother, why do these people want to go back?' When Oma Nath could not satisfy them they turned to Madan Singh and begged him 'If we have offended you in any way, punish us as you like, but please do not go back in this way. If you do, the whole village will be dishonoured.'

Madan Singh said only, 'Ask Oma Nath. He will tell you.'

Ever since Pandit Krishan Chandr had seen Sadan he was in seventh heaven. It was time for the main event and he was waiting for the bridegroom to come in. Instead, several people came in to tell him that the bridal party was preparing to leave. He asked them if there had been an altercation with Oma Nath. But they said, 'We don't know. But Oma Nath is with them, trying to placate them.'

Fuming, Krishan Chandr went in the direction of the tents. For the bridegroom's party to go back without solemnizing the marriage was no trivial occurrence. 'Do they think it is a doll's wedding?' he said to his companions, seething with rage. 'If they didn't want to carry it through why did they come? There will be bloodshed if they go back even if I have to go to the gallows for it!'

When he arrived at Madan Singh's tent he called out, 'Where is Madan Singh? Come out Maharaj!'

Madan Singh came out of the tent and said dryly, 'What do you want?'

Krishan Chandr: Why are you taking the bridal party back?

Madan Singh: Because I want to. We don't want to go through with this marriage.

Krishan Chandr: You will have to stay and carry it through. You can't go unless you do so.

Madan Singh: You can do what you like, but we are not going on with it.

Krishan Chandr: Why not?

Madan Singh: Don't you know why?

Krishan Chandr: If I knew it why would I ask you?

Madan Singh: Ask Pandit Oma Nath.

Krishan Chandr: I am asking you.

Madan Singh: I don't want to embarrass you, so don't ask me.

Krishan Chandr: I see. It's because I was in prison. How just you are!

Madan Singh: We wouldn't go back for that reason.

Krishan Chandr: Perhaps Oma Nath was niggardly with the dowry?

Madan Singh: We are not that mean!

Krishan Chandr: What other reason can there be?

Madan Singh: I am telling you, don't ask me!

Krishan Chandr: You will have to tell me. Do you think it is a trivial matter to come for the wedding and then go back without performing it? There will be bloodshed. Don't imagine you'll be let off easily!

Madan Singh: That doesn't worry us. We'd rather die than marry your daughter. We did not come here to stain our honour.

Krishan Chandr: So, do you think we are inferior?

Madan Singh: Yes, we think you are inferior.

Krishan Chandr: Can you prove that?

Madan Singh: Yes, we can.

Krishan Chandr: So why do you hesitate to tell us?

Madan Singh: All right, I'll tell you, and don't blame me. Your daughter Saman who is this girl's real sister, has become a prostitute. If you don't believe me you can go and see her in the red-light district.

Krishan Chandr said, 'This is completely false, a lie!' Then he remembered that when he had asked Oma Nath for Saman's address he had made some excuse not to tell him. Not only that, he suddenly understood the meaning of the many veiled taunts with which Janhvi frequently plied him. Convinced, he hung his head in humiliation. A moment later he had fainted and fallen to the ground. While all this was happening, hundreds of people were standing on both sides, but they were all struck dumb, to a man. The matter was so delicate that there was nothing to be said.

By midnight all the tents had been unpegged and the grove was once again dominated by darkness. The jackals had gathered once more and the owls were screeching their nightly song.

– 11 –

Bithal Das had admitted Saman in the Widows' Ashram secretly. None of the members of the committee knew about it. He had told the widows in the ashram that Saman was a widow too. But it was impossible to keep the news from privileged people like Munshi Abul Wafa. He managed to find Hira and got her to tell him Saman's whereabouts. He then informed his lecherous companions and the result was that these gentlemen became extremely solicitous about the ashram, so much so that one or the other would arrive there every day. The Seths, Chaman Lal and Balbahadar Das, Panditji, and Munshi Abul Wafa, became frequent visitors. These respectable people had developed an eager interest in the sanitation, interior decoration, finances, and management affairs of the ashram, and became very preoccupied with its welfare.

Bithal Das was deeply troubled. Frequently he would make up his mind to resign. He would say to himself, 'Am I the sole individual responsible for this ashram? There are many others in the committee who can look after it. They can do what they think is appropriate. At least I will not have to stand by and watch this degradation.' Sometimes he thought he should expose the spurious well-wishers and then weather the consequences. However, whenever he thought of the problem calmly, he felt that the ashram was tied up too closely with his own being, to be so easily severed. It was he who had founded it and it was to him that it owed its life. And if he were to abandon it, it would soon cease to exist.

So he ignored the fraudulent philanthropists, ridiculing their solicitude and refusing their unctuous suggestions. He would also hint to them that their constant hovering around the ashram was repugnant to him. But the self-centred are rarely sensitive. The two Seths would become courtesy itself whenever he held forth. Tiwari would behave as though he could never be provoked into anger. His good humour and sycophancy would in fact soften Bithal Das.

A month passed in this way. It was early morning one day, and Bithal Das was lost in his usual worries. He decided finally that he would get rid of his predicament and even if the ashram closed down it would not matter—such an end would be more desirable than its fate at the hands of these disreputable elements. Suddenly Bithal Das saw a phaeton stop at the door of the ashram. And who climbed out of it? None other than Abdul Latif and Abul Wafa.

Bithal Das became restive. He would have liked to drive them away, but was forced to show restraint.

Said Munshi Abul Wafa, 'Good morning, sir. You look a little worried. By God, it does one good to see such unselfishness, to know that there are still some beings who deserve to be called human. Lucky is the nation in which there are people like you. Our selfish, self-centred people have no taste for such qualities. Even those who have a pious reputation are not free from selfishness.'

Abdul Latif: Don't talk of our people! Call them selfish, self serving, unprincipled—and you'd still be sparing them. Even the best of them are hypocrites. But you are a being of infinite goodness. It seems as though God chose you from a group of angels and sent you down to this lucky nation.

Abul Wafa: Your virtuous soul has a fascination all its own. I would like to place an order for some embroidery. A friend of mine has asked for many dozens of embroidered bed sheets. I could order these from many other places in the city but I thought I should place the order at the ashram. If you have samples, could you show them to me, or if it is inconvenient for you now, I could come another time?

Abdul Latif: I too need many rolls of embroidered muslin for my family. The shops are full of such rolls from Lucknow, but I'd rather the benefit goes to the ashram.

Bithal Das responded dryly, 'We don't do embroidery at this ashram.'

Abul Wafa: But you should. Some of the ladies must be skilled at it. We are not in a hurry, we can come again, once, twice or even ten times. After all, if you can sacrifice so much, this is the least we can do. In such matters I don't believe in making religious distinctions.*

Bithal Das: I am very grateful to you for your concern, but the committee has decided that there will be no embroidery here because it weakens the eyes. So I cannot help you.

Saying this Bithal Das stood up so that the two gentlemen had no alternative but to leave. Cursing Bithal Das in their hearts, they left.

The sound of the phaeton had not yet died down when Chaman Lal's motor car pulled up. Sethji climbed out, supported by a stick. He shook hands with Bithal Das and said, 'What have you decided about the play? 'Shakuntala'* is a favourite with the English. If the ladies know some parts I would like to hear them rehearse.'

Sometimes, when we are faced with the need, we think of tactics which elude us when we are trying hard to find a strategy. Bithal Das had given a lot of thought to finding a way of getting rid of Chaman Lal, to no avail. But now suddenly a scheme took shape in his mind. He said, 'The play has been ruled out. We decided to consult the Collector, and he forbade it. I can't imagine what these people understand by 'politics.' Yesterday when I requested him for an annual donation for the ashram, he said, 'I cannot give aid to political* institutions.' I was stunned. When I asked him why he thought the ashram had a political character he said only that he couldn't answer that question.'

Seth Chaman Lal was quaking. He said, 'So the Sahib thinks that the ashram is political?'

Bithal Das: Yes, he said so clearly.

Chaman Lal: If he thinks so, they must be keeping an eye on all the people who come here?

Bithal Das: They would, yes. But that will not matter to those who are genuinely interested in the welfare of the people.

Chaman Lal: No, thank you. I am not one of those well-wishers. If someone told me that they consider Ramlila political, I'd even stop that. I am terrified of politics. You will not find a single volume of the Bhagvatgita* in my house. I have given strict instructions to my servants never to bring home their shopping wrapped in newspapers—they have been told to have the stuff wrapped in leaves. I don't allow the smallest bits of newspapers in my house. There was an old picture of Rana Partab Singh* in my room, I had it locked away in a trunk. I'll take my leave now.

Chaman Lal then leaped into his car, holding his enormous belly, and in a moment the car had disappeared amid a small storm of dust. Bithal Das was very amused and congratulated himself on thinking up such a good scheme. But it didn't occur to him that he had told a lie and had done some damage to his soul. This good man who abstained from all falsehood in his personal affairs, had no qualms in resorting to it when it was a question of public welfare.

After Chaman Lal had left, Bithal Das picked up the register in which he kept an account of the donations that he received, and was about to go out on a quest for more donations when he saw Seth Balbahadar Das arrive on a bullock cart.

A wave of anger ran through him. He threw down the register and sat down in a belligerent mood. There was no route of escape.

Balbahadar Das came up and said, 'I wonder if you planted the seedlings that I sent you yesterday. I'd like to see how they are doing. If need be I can send my gardener.'

Bithal Das replied dryly, 'No, thank you. You needn't send the gardener, nor can the seedlings be planted here.'

Balbahadar Das: Why can't they be planted? My gardener will do it for you. Have them planted today or they will dry up.

Bithal Das: They may dry up or not, but they can't be planted here.

Balbahadar Das: If you were not going to plant them why didn't you say so earlier? I sent for them from Saharanpur.

Bithal Das: They are lying in the verandah, take them back.

Sethji was a self respecting and outspoken man. Generally polite and amiable, he could turn into a demon of rage, if treated with the slightest condescension, which was the reason why he was considered haughty and bad tempered in certain circles. But at the same time he was much liked for his better qualities, by the people. They had complete confidence in him. They believed that he would never forsake the truth, that he would not let others down in order to further his own interests. They did not repose such confidence in Dr Shyama Charan. The ordinary man does not regard education, intelligence, and a prominent position in society, as highly as he does moral fibre.

Bithal Das's pigheadedness infuriated Balbahadar Das. He said, stiffening, 'Why do you sound so angry?'

Bithal Das: I can't be diplomatic.

Balbahadar Das: Nobody is asking you to be diplomatic, but you needn't be so brusque. It is discourteous.

Bithal Das: I am not going to take lessons in courtesy from you!

Balbahadar Das: I am a member of the committee, too.

Bithal Das: I know.

Balbahadar Das: And if I wanted to, I could have been its president.

Bithal Das: I know.

Balbahadar Das: My contributions are no less valuable than other people's.

Bithal Das: There is no need to bring up all that!

Balbahadar Das: I can destroy the ashram.

Bithal Das: Impossible.

Balbahadar Das: I can make the members of the working committee dance to my tune.

Bithal Das: Maybe.

Balbahadar Das: I can wipe out your ashram!

Bithal Das: Not possible.

Balbahadar Das: What makes you so audacious?

Bithal Das: My trust in God.

Glowering at the ashram, the Seth went back into the bullock cart. Bithal Das was not in the least put out by his threats because he felt certain that he would not carry them out. He would not slander the ashram, he thought, his pride would not allow him to stoop so low. He might even praise the ashram to the members, just to conceal his mortification. But at the same time, Bithal Das knew that the incident would not be forgotten and might well have unpleasant consequences, for hurt pride smoulders indefinitely and is never really put out. However, in spite of this realization Bithal Das did not feel the dejection that unpleasantness usually brings; instead, he felt the contentment of having done his duty, and in fact he was regretting not doing it earlier. Indeed he felt so serene that he began to sing aloud.

Meanwhile Padam Singh was coming to see him, looking worried, pale, and downcast in the extreme. It seemed as though he had just wiped his tears after a spell of weeping. Bithal Das came forward and embraced him saying, 'Are you ill? You look quite changed.'

Padam Singh: No, I am not ill; but I have been worried.

Bithal Das: The wedding went off well I trust?

Staring at the ceiling, Padam Singh replied, 'The wedding! All we did was to ruin the life of a poor girl. She turned out to be Saman's sister, and as soon as my brother heard this he decided to cancel the wedding, and took the bridal party back.'

Bithal Das: But this is a tragedy! Didn't you talk to your brother?

Padam Singh: Talk? I argued, even quarrelled with him, was even struck by him for my persistence, and I took that; but all to no avail.

Bithal Das: The poor girl! Saman will be miserable if she hears this.

Padam Singh: What's the news here? Did Saman's arrival cause a stir? Do the widows hate her?

Bithal Das: If the secret gets out, the ashram will be emptied.

Padam Singh: And how does Saman conduct herself?

Bithal Das: As though she had lived here all her life. It seems as though she wants to efface the blot on her life with her exemplary conduct now. She is always ready to do whatever needs to be done, and does it cheerfully. Even before the widows wake up in the morning, she finishes cleaning their rooms. She gives cooking lessons to many of them, and teaches many to sew. All of them consult her in all matters. She rules this establishment. It is beyond my expectations! She has even started reading. And although God alone knows the secrets of a person's heart, it seems as though she is a changed person.

Padam Singh: No sir, her character was never bad. She used to be a regular visitor to my house, for months, and my wife had a high opinion of her. Her misfortune was the result of some untoward circumstances. In truth, she had to suffer for our stupidities. And what about the rest? Has Balbahadar Das been up to his tricks?

Bithal Das: Of course. Do you think he can lie low for long? He is very active these days. Three days ago the Hindu members had a meeting. I couldn't go but I heard that their side triumphed. Now they have six votes including the president's two; while we have only four. Though, counting the Muslim votes we should have an equal number.

Padam Singh: So we need at least one more vote, to win. Is there any hope of getting it?

Bithal Das: I don't see any hope of it.

Padam Singh: Let's go see Doctor Sahib and Lala Bhagat Ram when you have the time.

Bithal Das: I am ready to go now.

– 12 –

Even though the doctor's bungalow was close by, the two gentlemen hired a vehicle to take them there. It was unfashionable to go to the doctor's on foot. On the way, Bithal Das gave his companion an exaggerated account of the day's happenings, and gave free expression to his views. Padam Singh said in a worried tone, 'Then we will have to become even more careful. Probably the whole weight of running the ashram will fall on us. Balbahadar Das may be silenced for now but his wrath is bound to make itself felt some day.'

Bithal Das: What can I do? I can't tolerate such unprincipled behaviour. These gentlemen are supposed to be models of the educated, civilized human being, yet their conduct is so abominable.

Padam Singh: Well, this was to be expected. It is the result of my own actions. More might transpire soon. Ever since the bridal party came back I have been extremely disturbed. I have no appetite and I toss and turn all night. The thought of that unfortunate girl never leaves me. If I start worrying about the ashram as well I will go crazy. I am stuck in a bog, and the more I try to climb out the more I am sucked in.

They arrived at Doctor Sahib's bungalow. It was ten o'clock and the doctor was sitting in his well-furnished room, playing chess with his daughter, Miss Kanti. Two terriers were sitting on the table and regarding the chessboard with concentration. Whenever they thought that a player had made a wrong move, they would put out their paws and change the position of a chess piece. Miss Kanti would laugh at this and say to them in English, 'Naughty.' On the left of the table Syed Tegh Ali was gracing a chair. He was advising Miss Kanti on her moves.

The two men entered the room. The doctor got up and shook hands with them warmly. Miss Kanti gave them a covert look and picking up a newspaper from the table began to read it.

Doctor Sahib said in English, 'Good to see you. Let me introduce you to Miss Kanti.'

On being introduced, Miss Kanti shook hands with the two men and said pleasantly, 'Papa was just talking about you two gentlemen. I am very pleased to make your acquaintance.'

Dr Shyama Charan: Miss Kanti has just come from Dalhousie. Her school closes in winter. They provide excellent education there, and she lives in the boarding house with English girls. The principal has praised her highly. Kanti, show them your principal's note. Mr Singh, you'd be amazed to hear Kanti's English speech; (laughing) she could teach me lots of English expressions.

Miss Kanti showed her certificate to Padam Singh, shyly. He read it and asked her, 'Do you study Latin too?'

The doctor said, 'She has been awarded a medal. Yesterday Kanti played such a hand at the club that the Englishwomen were taken by surprise. She gave them a run for their money! Now, why were you not present at the meeting of the Hindu members?'

Padam Singh: I had to go home.

Shyama Charan: Your proposal was being discussed. I think that you should not be in a hurry to place it before the Board. There is not much hope of success just yet.

Tegh Ali: You will get full cooperation from the Muslim members.

Shyama Charan: True, but the Hindu members are divided.

Padam Singh: With your help we are bound to succeed.

Shyama Charan: I sympathize with your views but as you know I became a member as a nominee of the government. So, unless I know whether the government approves of the proposal, I cannot give my views.

Bithal Das said crudely, 'If as a member you are not free to express your views, I think you should resign.'

The three men looked at him reproachfully. His comment was uncalled for. Tegh Ali said sarcastically, 'If he resigns, how will he be accorded the prestige he now has? How will he sit on the chair next to the governor's? How will he continue to be called "The Honourable..." How will he have the right to shake hands with the most distinguished Englishmen? How will he be entitled to get invitations for grand government functions? How will he be able to go to Nainital? How will he be able to make speeches? Just think!'

Bithal Das was crushed. Tegh Ali had shaken him up mercilessly. Padam Singh reproached himself for bringing such a tactless man with him.

Dr Shyama Charan said gravely, 'People imagine that we seek membership for the prestige it brings. They do not realize what an onerous responsibility it is, how much time, effort, and money the poor member has to expend. And what reward does he get except the satisfaction that he is working for his country and his people? If that satisfaction were not there, nobody would want to be a member.'

Tegh Ali: You are absolutely right. It is only those who have to shoulder the responsibility that know what a burden it is.

It was eleven o'clock. Padam Singh and Bithal Das left the doctor's house. Bithal Das said to Padam Singh, 'It is time for my lunch, so I'll go now. Come in the evening.'

Padam Singh replied, 'Yes, go by all means.' It occurred to him that if even a staunch man like Bithal Das could get restless when his meal was slightly delayed, what could be expected of others? People become public servants but are not ready to brook the slightest inconvenience. Thinking these thoughts, he arrived at Lala Bhagat Ram's house.

Lala Bhagat Ram was sitting on a wooden bench in the sun and smoking his hookah. His little daughter who was sitting in his lap kept leaping up to catch the smoke. On seeing Padam Singh, Lala Bhagat Ram stood up at once, and said after greeting him, 'I heard yesterday that you had returned. I thought I'd go and see you this morning but couldn't because there was some problem on my hands. A contractor's job is very troublesome; one has to get the work done, spend one's own money and yet one must wheedle others. Nowadays the engineer is displeased with me. He doesn't seem to like my work at all. My contract was to build a bridge, which I have had to pull down three times. He finds fault all the time. Not only have I made no profit, I am suffering losses all the time. But there is nobody to turn to. You must have heard about the proceedings of the Hindu members' meeting?'

Padam Singh: Yes I have, and I was very sorry to hear about it. I had pinned my hopes on you. Don't you agree with my proposal?

Bhagat Ram: I not only agree with it sir, I want to help you, with all my heart. But I am not the master of my own views. I've sold my freedom of opinion for my interests. I am like a gramophone record— I say what has been recorded.

Padam Singh: But do you agree that national interests should have precedence over personal concerns?

Bhagat Ram: Yes, sir, in principle I agree with it, but I don't dare put the principle into practice. As you know all my business depends on Seth Chaman Lal. If I annoy him I lose all my prestige. I am neither educated nor intelligent. I can't afford to lose my standing which is the only asset I have. Who would give me interest-free loans of thousands of rupees if I lost it? And then, it is not only myself that I am worried about, I need at least three hundred rupees per month to support my family. I am ready to make any personal sacrifice for the public good, but how can I drag my family into poverty?

When we are unwilling to do our duty we manage to give the most persuasive reasons for not doing it, in order to escape censure. At such times we abandon all self respect and reveal our private affairs, which we would otherwise never lay bare. Padam Singh understood that clearly Bhagat Ram could not be counted on. He said, 'In that case nobody can force you. I need just one more vote; tell me how I can get it.'

Bhagat Ram: My advice would be to see Kanwar Sahib. You can definitely get his vote. Seth Balbahadar Das has brought a suit of thirty thousand rupees against him, so Kanwar Sahib is mad at him these days. He would be glad to shoot him if he could. And yes, there is another way to persuade him: just make him president of some meeting, and he is your man.

Padam Singh said, laughing, 'Thanks. Yes, I'll go and see him.'

It was afternoon, but Padam Singh was neither hungry nor thirsty. He sat in his carriage and set forth. Kanwar Sahib lived in a bungalow by the Burna, so it took him half an hour to get there. No attempt had been made to clean or adorn the environs of the bungalow. There were no flowers and no greenery. In the verandah, several dogs were tied up with chains. Kanwar Sahib was fond of hunting and sometimes he went to Kashmir on hunting trips. At the moment he was playing the sitar in his room. The walls of his room were adorned with deer and leopard skins.

Several guns and spears stood in one corner, and on a large table at the opposite side of the room sat a crocodile. Entering the room, Padam Singh gave a start when he saw the reptile— the taxidermist had done a good job.

Kanwar Sahib embraced Padam Singh warmly, and said, 'One doesn't see you much anymore. When did you return from your home?'

Padam Singh: Yesterday.

Kanwar: You look pale. Have you been ill?

Padam Singh: No, it was nothing.

Kanwar: Have a cold drink, you must be thirsty.

Padam Singh: No, thank you. Don't trouble yourself. You've been practising the sitar?

Kanwar: Yes, I love my sitar. I am sick of the harmonium and the piano. Those English instruments have forced our music to take the back seat. It has lost its significance. And theatre has provided the finishing stroke. The only thing people like these days is the ghazal and the qawwali.* Our music will soon be extinct, like our martial arts. Without music, our hearts are no longer receptive to the exquisite, and this void shows up clearly in our literature. How unfortunate it is that the same people who produced a peerless work like the Ramayan now have to depend on translations even for light literature. In Bengal and the Deccan the tradition is still alive, so the people there are not so wanting in feeling. Anyway, so what seems to be the fate of your proposal?

Padam Singh: That is a cruel question. I expected more sympathy from you.

Kanwar Sahib laughed out aloud. His laugh echoed in the room. The brass shield on the wall vibrated with the sound of it. He said, 'You probably misunderstood me. In my speech I said everything in my power to support you; what else could I do? In fact I thought it useless to talk seriously with those who were opposing your scheme. Instead, I adopted a style of satire and ridicule; (remembers) ah, yes, I see (laughs aloud) if that is so I'd say that the Municipal Board is made up of fools. They probably didn't even understand my satire! The city of Benaras does not have a single discerning individual among the enlightened, cultured, and wise members of its Board! I am very sorry indeed that you misunderstood me. Please forgive me. I agree with your proposal, completely.

When Padam Singh left Kanwar Sahib's house he felt as refreshed as though he had been on a pleasant outing. His host's warmth and geniality had captivated him.

– 13 –

When Sadan returned home he felt like the man who comes home after many years, full of expectations and hopes; but when he opens his trunk he finds all his savings gone.

Independent thinking comes from education, company, and experience. Sadan was lacking in all these three prerequisites. He was at that stage of life when we are only proud of our religious and social customs; we can find no fault with them ourselves and do not have the courage to hear any criticism of them. At this stage of life we cannot discriminate between 'what' and 'why.' Sadan could run away from home himself but found it indecorous to take the women of his household to bathe in the Ganges. If women's laughter were heard in the men's part of the house, he would come in scowling and chide his mother. He regarded moral weaknesses not with the generosity of the philosopher but with the dryness of the ascetic. It is true that he loved Saman Bai, but in his religion this love was not as unforgivable a sin as bringing Saman or anybody connected to her, to his home. He had never eaten even a paan in Saman's house.* He considered the gentility, and ceremonial customs of his family more significant than the dictates of his conscience. The thought of the humiliation which a family connection with a fallen woman was bound to bring, was intolerable for him. Rather than that, he would have preferred to drown himself. When he heard Padam Singh advocate the completion of the marriage, he was incensed, and was only afraid that his father might give in to his uncle's persuasion. He couldn't understand why Padam Singh was supporting such a mad idea. Had anybody else suggested it he would have immediately silenced him. But since he had a great deal of regard for his uncle, the agitation in his heart found no expression, even though there was a storm of counter arguments in his mind. In fact he had never in his life been so rearing for an argument and it is possible that despite his inhibitions he would have locked horns with Padam Singh, had his sympathies not risen up and gathered round his uncle when Madan Singh dealt him the blow.

Having despaired of his marriage, Sadan's heart leaped once more in the direction of Saman. It is difficult to stop desire once it has been aroused. He went back to Benaras with Padam Singh, but once there, he found himself in a dilemma. It occurred to him that Saman would have got to know the whole business. She must have been disowned by her family so she couldn't have gone to the wedding, but it was unlikely that she had not heard of it. And if that were so, she might even decide to have nothing more to do with him.

But by the evening the optimism of desire had overcome his doubts, and Sadan set out for the red-light district. On the way he thought about what Saman might say to him and what he would say to her. 'What if she knows nothing about what happened? Then she might put her arms around me and say, "How faithless you are!" That would be so sweet!' The thought excited him so much that he spurred his horse to a gallop, and soon reached his destination. But once he was in the district, Sadan became like the child who on reaching the gate of his school, is reluctant to enter. He came and stood in a place which gave him a clear view of Saman's residence. And then he looked with nervousness at her door. It was locked. A burden seemed to have dropped from his heart. This disappointment was far more satisfying than success would have been. He felt the kind of happiness a man might feel who, despite having little money, takes his child to a toy shop at his insistence, but finds the shop closed.

But Sadan's satisfaction did not last very long. When he returned home he was sad and felt a void in his heart, as though something was lost. At night, after everybody had gone to bed, he got up quietly and again went to the red-light district. It was a winter's night, and the moon was peeping through the mist, running along fast like a nervous man. Sadan ran like the wind until he got to the district, but once there he was rooted to the ground, and all his excitement left him. He was thinking, 'It is shameful for me to come here at this time. What will Saman think? All the servants must be sleeping. And she will wonder why on earth I am here. How could I be so stupid!' And turned back immediately.

The next evening he again left for the red-light district. He had decided that only if Saman saw him and called him in would he visit her, but if she did not call him he would immediately go back. Her calling him in would be proof positive that she was not harbouring hard feelings against him, because why else should she call him in after the disastrous episode of his wedding? When he had gone a little further he thought, 'Would she be sitting on the balcony to call me? But how can she know that I am back? No, I should go and see her. Saman can never be angry with me for long. If she is still upset I will make my peace with her. I will do everything in my power to win her back, I'll beg her with folded hands, I'll fall at her feet, and I'll wash away her grief with my tears! No matter how weary she is of me, she can never forget my love. If only she would gaze at me once with those lotus-like eyes of hers, brimming with tears, I'd do anything for her! I could give my life to make her happy, so then, will she not forgive me?'

But as soon as he reached the red-light district his optimism and hopes vanished. He thought, 'What if she thinks when she sees me "There goes the self-styled Kanwar Sahib, as though he was really the master of an estate. What a hypocrite!" ' His feet became rooted to the ground at the thought and he was unable to go further.

Many days passed in this way. His hopes which were wont to rise during the day, crumbled in the evening when he approached the district where his beloved lived.

One day he wandered into Queen's Park. People were sitting on the ground under an awning, listening to Professor Romesh Dutt's forceful speech. Sadan dismounted from his horse and began to listen attentively. He came to the conclusion that prostitution was indeed a lethal poison for society, and that prostitutes should indeed be expelled from the city. If Saman had not lived there in the marketplace, he thought, he would not have been trapped in her love.

The next day he went again to Queen's Park. This time it was Munshi Abul Wafa who was holding forth. Sadan again listened carefully to the speech. He said to himself, 'It is true that these people are unfairly condemned. It is true that they keep the name of our deities alive. It is also a fact that only in their salons do Hindus and Muslims intermingle. Mutual rivalries and jealousies are unknown there. And it is there that people take refuge from the sorrows and disappointments of the world. Undoubtedly, to expel them would be cruel not only to them but also to the whole population.'

After several days his views underwent yet another change, and this state of affairs continued. He did not have the capacity to analyse a situation and form his own opinion, so he was swayed by every argument that was forcefully expressed.

One day he saw a notice announcing Padam Singh's speech. He began to get ready for it at three o'clock and reached Bini Gardens at four. Nobody had arrived there except the few people who were preparing the place for the meeting. Sadan joined those who were laying the rugs for the audience to sit on. By five o'clock a crowd had started assembling and in about another half an hour thousands of people had gathered. He saw Padam Singh arrive in a phaeton and his heart began to throb. The meeting began with a poem which was written for the occasion by Syed Tegh Ali and read out by Mr Rustum Bhai. After he sat down, Lala Bithal Das stood up to speak. Although his speech was dry and without any spark of liveliness, people were listening to him intently. His selfless work had won him their trust. They listened to his dry, colourless speech with the eagerness of a thirsty man drinking water. But the water he offered was more welcome than other people's sherbet.

Finally, Padam Singh stood up to speak. Sadan felt a kind of tickling in his chest, as though something extraordinary were about to happen. The speech was fascinating, and conveyed much sensibility. Its lucidity and refinement of language won over people's hearts. Now and then it became so moving that Sadan felt awe stirring in his every pore. Padam Singh was saying, 'We have not asked for their expulsion because we hate those women. We have no right to hate them—it would be a great injustice to do so. They are the victims of our lust, the weaknesses of our culture, our contemptible customs. They are in fact all of those evils in human form. The courtesans' quarter is a reflection of our blemished culture, it's the living picture of our demonic depravity. How can we bring ourselves to look down on them? Their condition is indeed pathetic. It is our duty to show them the right path, to reform them. And this will be possible only when they are relocated outside the city, beyond the reach of loathsome temptation.'

Sadan was listening closely to the speech. When the people next to him showed signs of appreciation, when clapping started spontaneously in the middle of the speech, Sadan was filled with happiness. But he was surprised to see many among the audience leave, one by one. Most of them had come to hear condemnation and abuse of the prostitutes. They found Padam Singh's liberal mindedness out of place. The public could understand the arguments of Bithal Das or Abul Wafa, but they found Padam Singh's wise approach incomprehensible. They could live on either bank of the river but refused to live in the middle of it.

– 14 –

S adan had acquired such a taste for speeches that whenever he heard of a forthcoming speech he made it a point to hear it. By listening carefully to dissenting views over a period of several months he developed the ability of forming opinions of his own. Now he was

no longer swayed by a novel argument, but looked for truth in a line
of reasoning. He began to realize that most speeches were merely a
collection of picturesque words, which were either entirely devoid of
merit, or were a rehash of old ideas. He developed the faculty of
analysis and began to think like his uncle.

But owing to his youth, Sadan's criticisms were prejudiced and
harsh. He did not have the maturity to perceive the sincerity of an
opponent's beliefs. He was sure that anybody who opposed his uncle's
view was himself a depraved person. He was so influenced by such
thoughts that he stopped visiting the red-light area. Now, whenever he
saw a prostitute out for an airing either in a phaeton in the park, or on
a stroll, he became indignant and had the urge to chase her away. He
would have liked very much to destroy all of the red-light district, and
as for those who watched the dance performances, and the women
who took part in them, he became convinced that they were the most
contemptible of creatures. Had he found them alone he would probably
have insulted them. And even though he still had some doubts in his
mind, none of them were about the salutary nature of the proposal. His
doubts he kept to himself lest they weakened the arguments for the
proposal. Saman still reigned in his heart, the desire to see her still
made him restless, her nut-brown beauty never lost its charm for him,
and so, to avoid thinking of her he stopped sitting alone, and always
kept himself occupied. He would go for a bath in the Ganges early in
the morning, read newspapers and books until ten o'clock at night, yet
despite all such restraints, Saman's memory would not leave him alone.
She would come to him in all kinds of disguises, sometimes cajoling
him and sometimes putting her arms around his neck. Then she would
give him a smile of love, and suddenly Sadan would be roused from
his daydream like one waking up from sleep; and brushing aside his
disquieting thoughts, he would begin to wonder why his uncle seemed
so sad—he hardly ever laughed these days. Why was Jitan bringing
him medicines every day? What could be the matter with him?
Suddenly Saman would enter his thoughts again and say with tears in
her eyes, 'Sadan, I did not expect this of you. You think I am a street
woman, but I have never been disloyal to you. I gave you all my love;
do you have no value for it?' Again Sadan would give a start and try
to brush away thoughts of her. He had heard in a speech that man is
himself the architect of his life. He becomes what he makes himself,
and the secret of keeping oneself unblemished is to stop dirty, immoral
thoughts from dominating one's mind. Sadan never let himself forget

this rule and he tried hard to suppress all such thoughts and fill his mind only with salutary and pious ones. In the same speech he had heard that it was not necessary to be highly educated in order to lead a blameless life; only pious thoughts and feeling were required. This made him seek purity of mind beyond everything else. Thousands of people had heard in the same speech that bad thoughts destroy not only this life but also the life to come. The smart ones had heard this and immediately forgotten it. But innocent Sadan had taken these words to heart—like a poor man who finds a gold coin, and values and keeps it as a great treasure. Those days Sadan was obsessed with the cleansing of his thoughts and if he chanced to look at a woman he would immediately reproach himself by saying, 'How can I ruin all my future life for the sake of one moment of pleasure in this earthly one!' Such self-admonition made him feel stronger.

One day, on his way to the river, he saw a procession of prostitutes in the marketplace. The most prominent among the city's prostitutes had held an urs,* and this procession was returning from there. Sadan had never seen such a manifestation of beauty, adornment, and animation. Such an alluring combination of silk, complexion, liveliness, elegance, and charm was an unusual sight for him, and though he tried very hard to restrain himself he could not help gazing at the beautiful forms.

He was like the student who, newly released from months of preparation for an examination, is immersed in recreation. One look was not enough, he looked again, and again, until his gaze was rooted in that direction as though somebody had chained it to the spot. He forgot to walk on and stood there, insensible and unmoving as a statue. He came to with a start after the procession had passed, and began to berate himself for ruining in a moment his months-long efforts. 'How weak I am,' he thought, 'and how self destructive!' But then he told himself that mere enjoyment of beauty could not be a sin. 'I did not look at them with evil intent, my heart was free from lust' he told himself, 'The innocent enjoyment of nature, which is the Gardener's handiwork, is a form of worship.'

Thinking these thoughts he walked on, but he was dissatisfied with himself. He told himself, 'I want to deceive myself. Why can't I simply acknowledge that I made a mistake. Yes, I have most certainly erred. But I consider my lapse to be forgivable. After all I am neither a saint nor an ascetic, I am merely a feeble-minded man. If I aspire to such a high standard, I will never be able to achieve it. Ah, but what a quality is Beauty! People say that sensuality drains the bloom on a face, but the sensuality of these women has in fact enhanced their beauty. They also say that the face is the mirror of the heart. That also is nonsense!'

To divert his own attention from himself, Sadan began to examine the subject from another angle. He thought, 'Yes, these women are beautiful in face and body. But how they have misused these blessings, how low they have fallen. They have bartered away a priceless asset like chastity for fine clothes and glittering jewellery. The eyes that should be showering rays of pure love, are brimming with mischief and sensuality. The hearts that should have been the fountainheads of love, are covered by poisonous dirt. What a tragic sight!'

These thoughts of loathing had a calming effect on Sadan. He strolled towards the river, but since his contemplations had made him late, he decided not to go to the bathing place where he usually bathed, since it was crowded by now. Instead, he went to the other bathing place adjoining the Widows' Ashram. This spot was usually deserted as it was further from town and therefore not frequented by people.

When he neared the bathing place, Sadan saw a woman there. He recognized her immediately. It was Saman. But how changed she was! The long black hair was not in evidence. The delicate body, the laughing, rose-like lips, the vulnerable, intoxicated eyes had all changed. She wore neither cosmetics nor jewellery and was clad only in a white saree. There was solemnity in her walk and on her face was an expression of sorrow and despair. The tale was the same but unornamented by metaphor; therefore it was simpler, and more moving. In his excitement at seeing her, Sadan at first walked fast, but when he saw the great change in her he fell back, as though it was not Saman but some other woman. The blaze of his love dimmed. He could not understand what had caused this change. He looked at her again. She was staring at him, but in her look there was listlessness instead of ardour, as though she had forgotten the past or wished to forget it, as though she did not want to fan the dying flame. It crossed his mind that she believed him selfish, hypocritical, and faithless. He looked at her once to see if his suspicion was correct. Their glances met and immediately parted. Sadan became sure that he had been right, and instantly pride awoke in his heart. He reprimanded himself for falling into the temptation of thinking foul thoughts once more, and he did not look at Saman again. She bowed her head and went away. Sadan saw that she was trembling, but he made no move in her direction. In his mind he had proved to her that if she had rebuffed him, he had spurned her too. But it did not occur to him that by not moving from his place he had given the impression of the very feelings that he wished to hide. When Saman was some distance away he began to follow her surreptitiously. He wanted to see where she was going. Desire had triumphed over resolve.

– 15 –

From the day the wedding party had gone back, Pandit Krishan Chandr took to his room. He was too humiliated to show his face to the world. He was altogether crushed by Saman's metamorphosis. The shame of it was more intolerable to him than that of his incarceration of three years. In prison it had calmed him to think that he was paying for his wrongdoing, but this new blow was playing havoc with his pride. He stopped seeing his disreputable companions who had shared his bouts of pot-smoking. He was aware that his stature had fallen even below theirs. He knew that everybody was gossiping about him. He thought people would be saying, 'This man's daughter...' and as soon as the thought crossed his mind, he cringed. 'Alas, if I had known that Saman would blacken the name of the family in this way, I would have strangled her. I know that she deserved to be the member of a noble family, that she loved the luxuries of life, but I had no idea that her conscience was so weak. After all, life has its ups and downs, even for the most fortunate. Everybody has his share of troubles. Women from the wealthiest families are known to have gone through times of abject poverty at some period of their lives, but have shown no trace of it on their faces. They might spend their day in weeping, but none would ever see their wet eyes. They die unburdened by other people's favours and they never lay bare their hearts to anybody. Those are the women! They live for the honour of their families and die for it. But this unfortunate, shameless woman! And what about her husband, why did he not decapitate her when she set foot outside her house, why did he not strangle her? He too must be shameless, immoral, unmanly. Had he been conscious of the honour of his family, matters would not have come to this. He is unabashed by his humiliation, but I am not, and Saman shall be punished. These hands that brought her up will lay the blade of the sword on her neck. These eyes that were refreshed by her charms, will find pleasure in seeing her writhe in blood. There is no other way to regain lost honour. The world will learn how those who are ready to die for their family's honour punish the shameless!'

Having determined on this lethal course of action, Krishan Chandr began to think of a way to execute his plan. In the prison he had picked up much information on how to commit a murder, for this used to be the most common topic of discussion. He decided that the best way to end his daughter's life would be to kill her with a sword. Once he had done that he would himself notify the police, and after that, he thought, the statement he would read before the magistrate would prove to be an eye opener!

Inflamed with the determination to shed blood, Krishan Chandr began to compose his statement. He was thinking, 'First of all I will mention the lasciviousness of civilized society. Next I will divulge the devious methods used by the police. Then, I will denounce and blast the tradition of giving dowries with such force that the audience will be thunderstruck. But the most momentous portion of my statement will be that in which I prove that the cause of our dishonour in fact lies in ourselves. Out of cowardice, fear for our own lives, fear of disgrace, because of false love for our children, our own shamelessness, our inability to protect our dignity; for all these reasons we try to cover up such immorality, with the result that base natures have become bold.'

Krishan Chandr had made his resolve without taking into account what would become of Shanta. His heart had been so filled by the sense of his dishonour that it had no place left for other considerations. He was like a man who leaves his child on his deathbed in his eagerness to wreak revenge on his enemy; who sitting on a dinghy, leaps at a snake in the water without giving a thought to the fact that his antics will overturn the dinghy.

It was evening. Krishan Chandr had decided to carry out his bloody design that day. He was feeling exhausted, with the dejection that soaks the heart just before one embarks on a dangerous mission. The fury and grief which had afflicted him for many days had given way to an all pervasive listlessness, like the aftermath of a high wind, when it slows down to a light breeze. Krishan Chandr was remembering the days when his life was free from trouble, when he would go out for a stroll in the evening accompanied with his two girls. He would pick up sometimes one and sometimes the other, and when they returned home how Gangajali would hug and kiss her daughters with delight. Happiness is more pleasant in the remembrance than in the experience. The same forests and mountains which seemed desolate and deserted, the same rivers and lakes which you passed by with blinkered eyes, will transform themselves into scenes of charm and beauty when called up by the eye of memory; and you yearn to revisit them. Krishan Chandr was overwhelmed by his memories, and tears streamed from his eyes. 'Alas, what a tragic end to a life that started so pleasantly! Here I am ready to kill my own daughter.' Suddenly he felt a wave of compassion for Saman. 'That poor girl has foolishly allowed herself to fall into a dark well, and should I now brutally throw stones at her?' But this leniency was short-lived. As soon as he remembered that Saman's door was now open for everybody—Hindus and Muslims alike could enter it at will, his outrage returned, and the fire of wrath raged in blinding radiance.

Meanwhile Pandit Oma Nath came and sat by him. He said, 'I've seen some lawyers. they have counselled filing a case.'

Krishan Chandr said with a start, 'What case?'

Oma Nath: Against those who took the wedding party back and called off the marriage.

Krishan Chandr: What good will it do?

Oma Nath: It will force them to either take the girl, or pay damages.

Krishan Chandr: But won't it make people talk more?

Oma Nath: We have lived through scandal. What more is there to be afraid of? I gave them one thousand rupees, spent four or five hundred on entertaining them, why should I give up all that money? With this amount I could find another husband for her. And in any case these enlightened gentlemen should be exposed.

Krishan Chandr said with a deep sigh, 'Give me poison before you file the case.'

'What are you afraid of?' Oma Nath enquired, irritably.

Krishan Chandr: Have you made up your mind to file the case?

Oma Nath: Yes, I have. All the prominent barristers and other lawyers of the city were gathered yesterday for the consultation. It is a novel case, and those people weighed its pros and cons very carefully before giving me their opinion. I've already engaged two lawyers and given them some advance on their fees.

Krishan Chandr said in a discouraged tone, 'All right, file the case.'

Oma Nath: Why does it grieve you so?

Krishan Chandr: If you don't understand, how can I tell you? If so far this matter has been talked about in a small area, litigation will make it famous all over the city. Saman is bound to be summoned for the court hearing. My name will be dragged in the mud.

Oma Nath: How long can I prevent that? After all, I have to marry off my own two girls. If I live with this stain on my reputation, won't it block their marriages?

Krishan Chandr: So you are filing this case in order to clear your own name?

Oma Nath said proudly, 'You can think so if you like. The bridal party went back from my door. People think Saman is my daughter. It is my reputation that is suffering everywhere. I'll put in a claim for ten thousand rupees, and even if we obtain a decree for five thousand, Shanta can be married into a good family. You know that it is necessary to sweeten damaged food before it can be eaten, so unless there is a bundle of money, how can Shanta be married? In a way it is my family's name that has been stained. Those who used to take pride in forming a relationship with me, will not speak to me unless their palms are oiled. This is how matters stand. I need money and there is no other way of finding so much.'

Krishan Chandr: All right, file the case.

After Oma Nath left, Krishan Chandr looked at the sky and said, 'O God, take me now. I cannot bear this disgrace anymore.' It was only now that he had become aware of the true extent of his dishonour. He realized that Saman's blood could not erase the stain of dishonour, in the same way that killing a snake will not detoxify its bite. What could be gained from murdering her but further notoriety? 'I would be arrested, and then after months of humiliation, I would be sent to the gallows. It would be far better to drown myself. Why not put out the light that shows only such fearful scenes? Alas, poor unfortunate Saman, she has pulled down with her, her helpless sister, ruined her life. My God, you alone are her guardian now. Only call me to you so that I may not have to witness her affliction!'

In a little while Shanta came in to call her father to dinner. Krishan Chandr had not seen her since the day of her aborted wedding. The glance he gave her now was full of great pain. In the dim light he could see an unnatural bloom on her face. Her eyes were filled with an expression of pure spirituality, and there was no trace of grief in her expression. Ever since she had seen Sadan she had felt in herself a pleasant strength. She had acquired a glorious self respect. Earlier, she barely spoke to her aunt; now she would press her legs for hours. She felt not a shred of jealousy or envy for her cousins. Now she would cheerfully fetch the water from the well, and actually enjoyed working the grindstone. Love had blossomed in her life. She had not found Sadan but had gained something far better, which was her love for Sadan.

Krishan Chandr was taken aback by Shanta's cheerfulness. It struck him that the blow had left dangerous traces.

He looked at her with guilt in his eyes, and said, 'Shanta?'

Shanta looked at him questioningly.

Krishan Chandr said tearfully, 'For the last four years the boat of my life has been caught in a maelstrom. In spite of it I had hoped that this difficult period would come to an end. But I can no longer watch the misfortunes of my children, and now I am going to jump from the boat into the waves. I know that all this has come about as the result of my own shortsightedness. If I had come to my senses earlier, matters would not have come to this. But what is the point of repenting now? If you ever meet the unfortunate Saman, tell her that I have forgiven her and that it is I who is to blame for her misdeeds. Two days ago I was ready to shed her blood but God has saved me from committing this sin. Tell her to be merciful to herself and her unfortunate parents.'

Krishan Chandr stopped speaking. Shanta was standing silently, she was sorry for her father. After a while Krishan Chandr spoke again. He said, 'Daughter, I want to make a request.'

Shanta: Tell me, I am listening.

Krishan Chandr: Just this, that you must never lose heart. Your fortitude will preserve you in the worst of times.

Shanta understood that he wished to go on but some scruple held him back. She knew what he was going to say, and she held her head proudly and looked at him. Her confident look told him all that, and much more than, the words that she could not bring herself to say.

– 16 –

After midnight, Krishan Chandr left the house. Nature, like an old woman, was lying idly, covered in a thick sheet of mist. On the sky, the moon with his face hidden, was running away, God knows where!

Krishan Chandr's heart was filled with a sudden longing to see Shanta again. She was the only token of his good days left in this world. In the unrelieved darkness of his despair she was the one faint light that was attracting him. He stood silently at the door for a while and then moved on with a sigh. It seemed to him that Gangajali was beckoning to him from the sky.

Krishan Chandr felt neither worry nor desire any more. He was tired of the world, and he wanted only to reach the river as soon as possible and be engulfed by its water. He was afraid that any delay might weaken his courage. To strengthen his resolve further, he began to run towards the river.

But soon he fell back and began to think. It was not difficult to jump into the water, after all. The ground would slip from under your feet, and that was the end! The thought sent a tremor through him. Suddenly it occurred to him that he could simply run away. 'If I don't live here, I won't be affected by scandal.' But he gave this thought short shrift. Greed for life could not ensnare him.

Although Krishan Chandr was not religiously inclined, the fear of the Unseen sent a shiver down his spine. To keep his resolve from faltering, he kept telling himself that God is forgiving and merciful. A sort of curtain had dropped on his innermost thoughts and he felt like a schoolboy who, after breaking the toys of a playfellow, is afraid to go back to his own home.

In this manner Krishan Chandr walked four miles in the direction of the river, his heartbeat quickening as his destination came closer. Fear had him in its grip, but he was trying to strengthen his dwindling resolve partly by reproaching himself for it and partly by walking faster. How shameless he was, he thought, to be afraid of death when there was nothing to live for!

Suddenly he heard the sound of singing, and as he advanced, the sound came nearer. The singer was coming towards him. In the silence of the night Krishan Chandr found the sound very melodious and he began to listen to it intently. Although the song was not particularly pleasing, it appealed to Krishan Chandr because the singer was competent, and he himself was a connoisseur of the art. The melody soothed his turbulent heart.

The song ceased, and a moment after Krishan Chandr saw a tall sadhu come towards him. The sadhu asked him his name and where he had come from, and on being told, asked him respectfully, 'Where are you going at this time?'

Krishan Chandr: There is something I need to do.

Sadhu: Why would you need to go to the Ganges at this time of night?

Krishan Chandr said sharply, 'You should be able to divine that, since you are a man of God.'

Sadhu: I am neither godly nor do I claim to be a sadhu. I am merely a Brahmin mendicant. But I will not let you go there at this time.

Krishan Chandr: Go your way. You've no right to stop me.

Sadhu: I wouldn't stop you if I had no right. You don't know me but I am your son-in-law. My name is Gujadhar Pande.

Krishan Chandr: I see. So you are Pandit Gujadhar Parshad. When did you take on this character? I have wanted to meet you; there is much that I need to ask you.

Sadhu: These days my camp is under a tree by the river. Come and rest there for a while and I'll tell you everything.

The two men were quiet, on the way. Soon they reached the tree under which the sadhu lived, and where a fire was burning. Some straw had been spread on the ground and on it were laid a deerskin for prayers, a water container, and a bundle of books.

Warming his hands on the fire, Krishan Chandr enquired, 'You are a sadhu now. Tell me honestly what happened to Saman.'

Gujanand was studying Krishan Chandr's face in the firelight. He could see on it his innermost thoughts written in bold. The sadhu was now no longer Gujadhar Pandit. The society of fakirs and meditation had illumined his inner being. The more he thought about the past the more he regretted it. He now felt a sympathy for Saman, and sometimes he felt like asking for her forgiveness. He said to Krishan Chandr, 'It all happened because of my stupidity. It was the consequence of my cruelty and brutal treatment of her. That woman was adorable. She deserved to be the mistress of a great house. A man as short-sighted, base, and undiscerning as I, did not deserve her. I was not capable of appreciating her fine qualities at that time. She had to undergo every physical hardship while she was with me, but she put up with it all cheerfully. She respected me, but I was suspicious of her. I thought she was faithless. I misunderstood her contentment, her calm, and her loyalty. I thought she was deceiving me. If she had fought with me, taunted me, or cried, I would have trusted her more. But it was her high-mindedness that caused me to think that she was unfaithful to me. Matters came to such a head that one day, just because she came back late from a friend's house, I turned her out.'

Krishan Chandr interrupted him to say, 'Didn't it occur to you that your cruelty was destroying a noble family?'

Gujadhar: Maharaj, how can I tell you what had come over me? But I can swear that she was innocent. She lives in the Widows' Ashram now and is respected by everybody there. They are all impressed by her decency.'

On hearing Gujanand's account Krishan Chandr had softened towards his daughter. But like a stream of water which flows in a different direction when it is stopped from its natural course, his anger, diverted from Saman, became directed at Gujanand. He looked at him wrathfully and said, 'You humiliated me, you ruined my daughter's life, and yet you sit before me as though you were some great saint. You should go and drown yourself.'

Gujanand was scratching the earth. He did not lift his head. Krishan Chandr spoke again, 'It was not your fault that you were poor. If you could not provide appropriately for your wife, that again was not your fault. That you could not understand her, not know her thoughts, I do not blame you for all of this; but why did you turn her out? Why didn't you kill her? If you had doubts about her chastity why didn't you cut off her head? And if you did not have the courage to do so, why didn't you kill yourself? Why didn't you take poison? If you had ended her life she would not have turned to what was worse, the reputation of my family would not have been stained. And you claim to be a man! Your shamelessness and your cowardice are contemptible. A man whose blood does not boil at his wife's infidelity is worse than an animal.'

Gujanand realized that in his enthusiasm for proving Saman innocent, he had got himself stuck in a quagmire. He began to regret his excessive and uncalled for magnanimity. He did not think that he deserved such a harsh judgement. A wounded heart seeks reproach that is mixed with sympathy. It does not look for severity and humiliation. A suppurated boil needs a cut from a knife, not a blow from a rock. He regretted his remorse, and once again became eager to pin the blame on Saman.

Suddenly Krishan Chandr thundered, 'Why didn't you kill her?'

Gujanand replied patiently, 'My heart was not so hard.'

Krishan Chandr: Then why did you turn her out?

Gujanand: Only because there was no other way of getting rid of her.

Krishan Chandr said mockingly, 'You could have poisoned yourself.'

Mortified by the taunt, Gujanand said, 'It is useless to end one's life.'

Krishan Chandr: A useless death is better than a useless life.

Gujanand: You cannot call my life useless.

Krishan Chandr: Is that why you are playing this part?

Gujanand: No, but for the reason that my life is of value to some people. Perhaps Pandit Oma Nath told you that I gave him fifteen hundred rupees for Shanta's marriage, which I had collected by begging. I have collected one thousand rupees more which I am going to give him.

Gujanand stopped as he realized that it was tasteless of him to have mentioned this, and bent his head in embarrassment.

Krishan Chandr said doubtingly, 'He never told me this.'

Gujanand: It was not a matter that needed to be brought up. I was wrong to mention it. Pardon me. All I meant to say was that I couldn't have done anybody any good by taking my own life. This blow made me work for the improvement of life. Our mistakes are a divine call that alerts our sleeping consciences for ever. Education, company, inducement, none of these have an effect as salutary as the consequences of our mistakes. Perhaps you are putting it all down to my insensitiveness to questions of personal honour. But this very 'insensitiveness' has become the source of my peace of mind and altruism. By ruining the life of one woman I am enabled to save hundreds of unfortunate unmarried girls from the fate of lifelong spinsterhood, and I am very happy to note that Saman's life too has been similarly blessed. Sitting in my hut I have seen her take sacred immersion in the holy Ganges, and I was amazed at her sincerity. Her face was lit by the light of her inner purity. She used to be accomplished in household affairs; now she is adorned with the beauty of her inner self. I have no doubt that one day she will be the jewel of womanhood.

At first Krishan Chandr listened as an intelligent customer might listen to the persuasions of a glib salesman—not forgetting that it was in the salesman's interest to convince him. But gradually he began to be moved by Gujadhar's words. He felt that he had spoken harshly to a man who not only reproached himself for his past actions but who was also his benefactor. 'How ungrateful I am!' he thought while his eyes filled with tears. A guileless man is like wax, in that it melts as fast as it hardens.

Gujanand looked compassionately at him and said, 'Be the guest of a fakir tonight. You can go where you like tomorrow morning and I'll go with you. Cover yourself with this blanket.'

'I don't need the blanket, I'll lie down without it.' Krishan Chandr replied mildly.

Gujanand: You think it would be sinful to use the blanket; but I don't use it, I've kept it specifically for guests.

Krishan Chandr did not protest anymore. He was cold, and fell asleep as soon as he had covered himself with the blanket. However it was not a very sound sleep, just a reflection of his heartache that came to him. He dreamt that he was lying on a bed in the prison and the jailer was giving him a look of hate. He was saying, 'You cannot be set free yet.' Meanwhile Gangajali and his father came and stood by his bedside. Their faces were blackened. Gangajali said, weeping, 'It is you who has done this to me.' His father said wrathfully, 'Is your disgrace to be the reward we get for our lives? Is this why we brought you into this world? Our faces will always remain darkened with the ignominy, we will always suffer this punishment. You called down this everlasting punishment upon us for the sake of your mortal existence, so now we are going to put an end to your life.' Saying this his father rushed at him with an axe.

Krishan Chandr woke up. His breast was trembling. Before he went to sleep he had forgotten the purpose for which he had left the house. The dream reminded him of it, and reproaching himself for forgetting it he became certain that it was not a dream but a divine reminder. Gradually, Gujanand's words of consolation faded from his memory, and he thought, 'Saman may now be the goddess of chastity, but nothing can remove the stain with which she has blackened our face. The holy man says that sin reforms the sinner, but I find this very unlikely. I have sinned many times but my sins have had no such reformatory effect on me. That was just his rhetoric, entirely absurd! He probably said it to cover up his own brazenness. Sin can only produce more sin. If goodness could come from sin there would have been no more sinners in this world.' Krishan Chandr had decided to put an end to his misery once and for ever.

The moon had set. The deepening mist had blotted out all distinction between hill and tree, bank and river. Krishan Chandr was walking on a narrow path, picking his way more by guesswork than by sight. He was so engrossed in avoiding the sharp pebbles and bushes in his path that he had stopped thinking of his predicament.

When he reached the river he could see some light. He went down to the river. Wrapped in a sheet of mist, the Ganges was moaning like a sick man. The only difference between the river and the surrounding darkness was the flow of the water. One of them was flowing darkness. All around there was the sadness which settles on a house after a death.

Krishan Chandr stood by the river. He was thinking, 'How close at hand is my end. In a moment who knows where this life will be! And what will happen to it there? Today my ties with this world will break. Oh God, be merciful, protect me!'

Then, for another moment he stood there hardening his heart. When he became certain that he was no longer afraid, he went into the water. The water was cold and his limbs were chilled. But not caring, he went deeper into the water. When it came up to his neck he looked again at the all-pervasive darkness. This was the last thread that linked him to the world, the last test of will, pride, honour.

All that he had been through so far was merely a preparation for the test that was coming. It was the last round between Will and Desire. Desire pulled him to her with all her force—Saman came to him in the garb of a holy ascetic, an anguished Shanta stood before him. 'Nothing is lost yet. Why not become a sadhu? I am not so well-known that the world will always drag my name through the mud. There are so many girls who fall victim to a life of lust, every day. The world does not care about them. I am a fool to think that I will be ridiculed.'

Will tried her best to counter this last argument, but was not successful. The distance between life and death was now no more than one plunge, one step. the step backward was so easy to take, the step forward so difficult, so daunting! Krishan Chandr lifted his foot to take a step back—Desire was showing its prowess, but in reality it was not the desire for life, it was the fear of the Unknown.

At that moment Krishan Chandr realized that he could not turn back. He was being dragged forward gradually. He screamed, he tried with all his might to move his chilled legs backwards, but the Moving Finger had written, and he was carried forward.

Suddenly he heard Gujanand call. Krishan Chandr answered as loudly as he could but before he had completed what he had to say, he was overpowered by the waves, like a candle blown out in the wind and drowned in the darkness. The fires of grief and pain were extinguished by the cold water.

Gujanand had heard just these words, 'I am drowning', and then he could hear no more beyond the tumult of the pitiless waves. He stood by the river for a long time. The same three words kept coming to him from all four directions. From the hills close by, from the waves before him and from the darkness all round, the words came echoing to him.

– 17 –

Early the next morning the news of the tragedy spread in Amola, but except for a few people none came to condole with Oma Nath. Had it been a natural death even his enemies would have arrived to commiserate, but suicide is terrifying and on this occasion friends were no better than enemies.

When Gujanand told Oma Nath the news, Oma Nath was bathing by the well. He was neither grieved nor surprised. Instead, he was annoyed at Krishan Chandr. The thought of police intervention had dismissed grief. That day he took very long over his bath. The anxious temperament is so engrossed by the contemplation of its circumstances that it loses all sense of time.

Janhvi raised a storm of lamentation. Seeing her wail, her daughters began to sob. Women neighbours arrived dutifully to commiserate. They had no fear of the police but the wailing did not last long. Instead, Krishan Chandr's strengths and shortcomings were analysed. The consensus was that his better qualities had been predominant. In the afternoon when Oma Nath came in to have a cooling drink and said some unpleasant things about Krishan Chandr, Janhvi gave him a sharp look and said, 'How tasteless to say such things.' Oma Nath was abashed.

Janhvi was secretly savouring her satisfaction. But she considered this so shameful that she kept it even from Oma Nath. Shanta alone was truly grieved. Even though she saw her father as dependent on others, he was the only being in the world who she could look upon as some kind of support. In fact it was his wretchedness that made her care for him all the more. Now she was truly alone in the world. But her affliction did not weaken her resolve, and she became even more compassionate. Her fortitude and patience became phenomenal. Like the last drops of the rainy season, the dying words of counsel of a person are never wasted. Now, she never said anything that might wound the spirit of her father. While he was alive she was sometimes short with him; but now she never allowed a selfish thought with regard to him to enter her head. She believed that once freed from the bonds of physical existence, the spirit becomes cognizant of the internal as well as external reality, in other words, that her father would now know her innermost thoughts.

Even though she spared no trouble to please Janhvi, the older woman taunted her frequently. It made her angry, but she swallowed her wrath and pride, and never gave vent to tears even in solitude. She was afraid that her father would be grieved by her weeping. For the festival of Holi, Oma Nath brought beautiful sarees for his daughters. Janhvi also wore a silk saree, but Shanta had to wear her old cotton saree. Her heart was overwhelmed with sadness yet there was no frown on her forehead. The two sisters had pulled long faces because their sarees had not been lined, while Shanta was discharging her household duties cheerfully. Even Janhvi felt sorry for her and pulling out an old silk saree, gave it to her to wear. Shanta did not spurn it; she put it on and went back to her work in the kitchen.

One day Shanta forgot to wash Oma Nath's loincloth. Early the next morning when Oma Nath was about to take his bath, the loincloth was found to be wet from the previous day's ablutions. Oma Nath said nothing, but Janhvi cursed Shanta so much that the girl began to weep. As she struck the cloth against the stone to clean it, she sobbed. Oma Nath was grieved to see this, and he thought, 'We are harassing an orphan for the bread that we give her. How can we show our faces to God?' He did not bring up the matter with Janhvi but in his heart he decided that he would soon put an end to this tormenting of an orphan.

As soon as the rituals for the dead were over, Oma Nath had busied himself with his plans to take legal action against Pandit Madan Singh. Lawyers had convinced him that he would win the case, and the hope of securing five thousand rupees had stirred him to make ambitious plans. This fresh preoccupation delighted him. A new house had been designed, for which the search for an appropriate plot of land was on. Engaged in these pleasant schemes he had forgotten to worry about Shanta. But Janhvi's harshness with her that day had reawakened his concern for her. The thousand rupees that Gujanand had given him, which he had put away for legal expenses were lying in the house. One day he spoke to Janhvi about Shanta's marriage. Shanta overheard this conversation. The talk about filing a suit used to depress her but she thought it entirely inappropriate to interfere. However the news that another marriage for her was being contemplated drove her to speak up. A wave of suppressed emotion banished all inhibitions, and as soon as Oma Nath left the house she went up to Janhvi and enquired, 'What was Uncle telling you?'

Janhvi replied in a dispirited tone, 'What can he say except bewail his luck? If that cursed Saman had not started the whole problem, he would not have had to repeat his efforts. Now, neither a good family nor a good man can be found. There is a village nearby where he has gone to look for a suitable match.'

Shanta said, staring at the floor, 'Am I such a burden for you that I have to be got rid of? Tell Uncle that he should not trouble himself anymore.'

Janhvi: You are his beloved niece. He cannot bear to watch your troubles and do nothing. I too had told him to let the matter be, for now, and think about the marriage after the money from the lawsuit had come in. But he doesn't listen to me.

Shanta: Why don't you send me there?

'Where?' Janhvi asked with surprise.

Shanta replied simply, 'Either to Chanar or to Benaras.'

Janhvi: You are talking like a fool! If that were possible there would be no problem. If they were willing to have you why would they have committed this outrage?

Shanta: If they don't want me as a daughter-in-law, they might keep me as a slave.

Janhvi said heartlessly, 'Why don't you go then? Nobody is stopping you. Your uncle will never agree to take you—only to have to bring you back later, with even more dishonour. He will crush them to make them pay damages.

Shanta: Aunt, no matter how proud they are, they will have pity on me if I go and stand at their door. I am sure that they will not turn me away. In that situation it is difficult to rebuff even one's enemy, while I...

Janhvi lost all patience. She could not tolerate such shamelessness. 'Be quiet!' she said, interrupting the girl, 'You don't have a shred of shame or dignity in you. Do you want to force yourself on them? If somebody didn't want me, I wouldn't look at him no matter how rich he were. If those people came now to beg us for you, I'd chase them away.'

Shanta said no more. No matter what the world thought, she considered herself married. For a girl who was engaged to one man, to be married to another was a repulsive idea. She had heard so much about Sadan's charms during the month preceding the arrival of the bridal party that she had lost her heart to him. During the duvar puja at her door she had looked at Sadan the way a woman looks at her husband, not as though he were a stranger. The thought of another man entering her life was like a stone hurled at the glass of her chastity. After thinking of him as her husband for so long she could not banish him from her heart. She could not break the chain of her conviction. Sadan was now her husband, whether or not he acknowledged her, whether or not he showed concern for her. If Sadan had come to her immediately after the duvar puja she would have treated him as a husband. A marriage is not just a ceremony, it is a perception of the heart.

Shanta still hoped to go to her husband's home some day. It agonized her to hear that another marriage was being proposed for her. She had brushed away her inhibitions to beg Janhvi to let her go to those whom she considered her in-laws—that was as far as she could let her thoughts run. But Janhvi's callous retort unnerved her. At night, after everybody had gone to bed she began to write a letter to Padam Singh. This was her last stratagem. If it failed, she knew what she was going to do.

It did not take her long to write the letter. She had already composed it in her mind, it just needed to be put down on paper:

> My respected father-in-law,
> I am in great trouble. Please have pity on me. Father was drowned in the Ganges, and they are preparing to start legal proceedings against you. They have also decided to marry me to somebody else. Please get in touch soon. I'll wait to hear from you for a week. After that you will not hear from this helpless orphan again.

– 18 –

Padam Singh was married for the first time when he was in college. By the time he passed the FA exam, he had fathered a son. But his inexperienced wife knew nothing about nurturing a child, with the result that the child who was born healthy, gradually became weak and within six months both mother and son passed away. Padam Singh had decided not to marry again, but on completing his course in law he felt compelled to remarry. It was seven years since he had been married to Sobhdra.

For the first two or three years after his marriage, Padam Singh was not troubled by his childless state. Whenever Bhama brought up the issue he would tell her that he felt no desire to have children, that he was not ready to shoulder such a responsibility. But when the fourth year came to an end and he was still issueless, he began to despair. As the days passed, his anxiety grew. He began to feel a sort of vacuum in his life and his love for Sobhdra diminished. Sobhdra fathomed his feelings, and was pained. But accepting the situation as her fate, she reconciled herself to it.

Padam Singh often told himself, 'What use is a child? For twenty-five years after its birth one must feed, clothe, and educate him, and even then one must worry about whether he will turn out well. If the child dies, it is a tragedy; and if one dies oneself the poor child's life is ruined. Who wants such a blessing?' But such thoughts did not console him. He tried to hide this feeling of deprivation from Sobhdra, and he wanted to continue loving her as before, but when the heart is wrapped in the darkness of despair where can the face receive the light of happiness? Even an unperceptive observer could sense the tension between husband and wife.

Sobhdra gave her husband a great deal of love and tried to please him so that he would forget his longing for a child. But in this difficult task she was no more successful than somebody who wants to cure a sick man with sweets. Even in the minutest of household affairs she had to give in to him. Ever since Sadan had come to stay, she had to put up with her husband's chiding on his account on numerous occasions. A woman can endure a wound inflicted by her husband, but if he gives her so much as a sharp look on another's account, she finds it intolerable. Sadan was a thorn in her flesh, and finally one day her frustration with him boiled over.

The cook had not come for some reason, and Sobhdra had to prepare the meal. She made fine chappatis for Padam Singh, but in her hurry to leave the kitchen, in the blistering heat, the chappatis she made for Sadan were thick and rough. When Padam Singh sat down to eat, he noticed the thick bread in Sadan's plate. Angry at this discrimination, he immediately exchanged his own carefully prepared chappatis with Sadan's. Sobhdra responded with some bitter taunts, and Padam Singh retaliated, blow for blow. This unpleasant exchange went on for some time until Padam Singh got up and left in a temper, leaving his food untouched. Sobhdra made no attempt at reconciliation, she merely cleared the table, and went and lay down. A whole day passed but neither of the two showed signs of softening. Neither ate when the cook prepared the meal the next day. Sadan went to both, by turns, begging them to eat, but the response from one was, 'I am not hungry', while the other said, 'I'll eat later', and, 'of course I'll eat. If I could have done without food I wouldn't have taken such bullying.' The surprising thing was that Sobhdra was talking affably to Sadan, despite the fact that he was the cause of all the unpleasantness the day before. The deer knows well that the arrow that comes from behind the screen is the hunter's love of hunting or his desire for meat.

Late in the afternoon, Padam Singh woke up and stretched himself. His heart was filled with ill feeling against Sobhdra. The postman brought a letter, postage unpaid. Padam Singh looked at the man acidly as though he had sinned by bringing such a letter. At first he thought of returning it unopened—it was bound to have come from a hard up client who had detailed his troubles in it. But changing his mind, he took the letter and read it. It was Shanta's letter. Having read it he put it down, then read it again and began to pace the room. Had Madan Singh been there, he would have shown him the letter and said, 'This here is the result of your fear of scandal, your pride in family dignity. You have murdered a man, his blood is on your hands.' He was quite pleased to hear about the filing of the suit against his brother. 'It would be a good thing if they file this suit,' he thought. 'It would put an end to his notions of gentility. And he will certainly have to pay the damages. My brother will then learn how costly was the spectacle he created. Alas! how this poor girl must have suffered.' Padam Singh read the letter again. Every word of it was a testimony to good faith. Shanta had called him 'Father-in-law.' This word worked on him like magic. It had provoked his sympathy and stimulated his desire to see justice done. He changed his clothes and immediately set out to see Bithal Das. When he arrived at his house he was told that Bithal Das had gone to Kanwar Anarudh Singh's. He at once turned his bicycle in the direction of Kanwar Sahib's house. There was no time to be lost. He was afraid that delay might cool his resolve.

At Kanwar Sahib's, a musician had arrived from Gwalior. Kanwar Sahib had invited his friends to listen to him. Padam Singh found that a dispute was going on between Bithal Das and Professor Romesh Dutt, while Kanwar Sahib, Pandit Parbhakar Rao, and Syed Tegh Ali were enjoying the dispute as though they were watching a cockfight. As soon as Kanwar Sahib saw Padam Singh, he greeted him and said,

'There is a bloody war going on here. Try and separate them, otherwise they will die fighting.'

Professor Dutt was saying, 'It is not a sin to be a theosophist.* I am a theosophist and the whole world knows it. It is thanks to our Society that today, in countries like America, Germany, and Russia, you can find people who believe in your Ram and Krishan, and enjoy reading holy books like Gita and Upanishads. Our Society has brought prestige to the Hindu nation and widened its influence. It has put the Hindu nation back on the throne of honour which had been lost to it for centuries due to its idleness and inertness. The Hindu nation would be guilty of ingratitude if it did not thank the people who brought light back to it with their torches. Whether these torches were lighted by Madam Blavatsky* or Col. Olcott,* our duty lies in showing gratitude to our benefactors. If you call this spiritual slavery, it is arrant injustice on your part.'

Bithal Das listened to this speech as though it were meaningless nonsense, and said, 'I believe that what you call gratitude is no more than spiritual slavery. In fact, even a slave is free because his spirit is unfettered, though his body may be in chains. But you have sold the very freedom of your spirit. Your Western education has made you so timid that even in your religious and spiritual beliefs you wait for the approval of European scholars. You do not respect the Upanishads because they deserve to be respected but because Blavatsky has praised them. You used to find your religious rites and customs quite meaningless but now that Westerners have propagated their good qualities, you are convinced of their excellence. You have lost the ability to use your own mind and judgement. Until just a few years ago you never paid any attention to Tantric vidhya.* But ever since European scholars revealed its meaning, you have become convinced of its efficacy. This intellectual submission is far more despicable than physical compliance. You read the Upanishads* in English and the Gita* in German, you say Arjuna for Arjun, Krishna for Krishan, and Rama for Ram; this is how well you know your language! You have, by reason of your intellectual slavery, accepted submissiveness as your fate in a country where we could have held our heads high, thanks to the knowledge and skills of our ancestors.'

Romesh Dutt's face went red. He was about to reply when Kanwar Sahib spoke out, 'Friends, I have to say this: Babu Bithal Das, you have to take back your accusation about slavery.'

Bithal Das: Why should I take it back?

Kanwar Sahib: You have no right to make this accusation.

Bithal Das: I don't understand you.

Kanwar Sahib: I mean that none of us has the right to call another a slave. In a gathering of the blind, which man has the right to call others blind? Each one of us, rich or poor, a raja or a fakir, is a slave. If we are ignorant, poor, or uncultured, we are less of a slave because we remember our Ram, we wear our loincloth and our traditional headdress, speak in our own language, rear our cow, and bathe in our sacred Ganges. And if we are educated, rich, and enlightened, we are more of a slave because we wear Western dress, speak a foreign language, bathe in a bathtub, and look down on our brothers. The whole of our nation is divided into these two groups, therefore not one of us can call another slave. It is incorrect to divide slavery into sub-types. Slavery is only of the spirit; all other types stem from it. Motor cars, bungalows, polo, and pianos, all these are iron chains. Those who wear them can never enjoy real freedom. Do you know those people who earn their bread with the sweat of their brow, who do not rely upon strangers for their language and their culture? The enlightened among us wear iron shackles, lose our spiritual freedom, and then look down on the farmers, thinking they deserve our pity. But it is we ourselves who are pitiable, who are dependent on others for our bread, who humble ourselves at the bungalows of our masters, the Westerners, who bear with the caprices of our cooks, run around clutching offerings of flowers. Has anybody ever seen farmers behave in such an undignified, absurd manner? We are tame dogs that hunt the free animals of the forest. There can be no better analogy for us.

Parbhakar Rao said, smiling, 'You should become a farmer.'

Kanwar Sahib: If I became a farmer how would I suffer retribution for my first existence? How would I celebrate Christmas? How would I bribe cooks to elicit their salute? How would I go on pilgrimages to Nainital in order to secure awards? How would I throw dinner parties and take the ladies' dogs on my lap? How would I oppose schemes for national welfare in order to please my masters? All these are the last stages of human baseness. Unless we cross this stage we can never be redeemed.

Kanwar Sahib then turned to Padam Singh and said, 'Tell me, when is your proposal coming up before the Board? You look a bit depressed these days. Is your proposal likely to meet the same fate as that suffered by most of our national enterprises?'

Over the last few days Padam Singh had in fact lost heart over the scheme. As the day set for the presentation of the proposal before the Board drew nearer, his confidence grew thin. He felt that it might do more harm than good. But he did not have the courage to make his doubts known. So he looked at Kanwar Sahib confidently and said, 'No, it's not that. But I have not had much time recently, and so have been unable to give it the same attention.'

Kanwar Sahib: There is no hitch, is there?

Padam Singh said, looking at Tegh Ali, 'We are depending on the Muslim members.'

Tegh Ali said, sounding profound, 'Relying on them is like building a sandcastle. You don't know their intrigues. Don't be surprised if they leave you in the lurch at the last moment.'

Padam Singh: I do not expect this of them.

Tegh Ali: That is your good faith. At the moment they are busy with the Urdu-Hindi dispute. Killing of cows, separate electorates, the proposal of a law on usury—they are devising ways to use all these issues to stir up religious prejudice.

Parbhakar Rao: Isn't Seth Balbahadar Das coming here? He should be brought here, somehow.

Kanwar Sahib: I didn't invite him because I knew he wouldn't come. A different point of view from his own is anathema to him. Most of our leaders have this attitude. This is the one point over which they become quite lively. Oppose them even slightly and they will be asking for your blood. Not only will they stop speaking to you and avoid you, they will even denounce you to the authorities, and ridicule you in their own circle. If you are a Brahmin they will call you a beggar, if you are a Khatri your appellation will be 'ignorant boor', if you are a Vesh you will be known as a petty shopkeeper, and if you are a Shudar* you are of course the basest of beings. If you are fond of singing you are a hypocrite and even your womenfolk will not be spared. In our land, holding a different opinion is the worst of sins, one from which you can never be redeemed. Aha! Look, there is Dr Shyama Charan's car.

Doctor Sahib dismounted from his car and said to the gathering, patronizingly, in English 'I am sorry I was late.'

Kanwar Sahib greeted him, and the others too shook hands with him. Then Doctor Sahib sat down on a chair and said, again in English, 'When is the performance going to begin?'

Kanwar Sahib: Doctor Sahib, you forget that this is a gathering of black people.

Doctor Sahib said with a laugh, 'Pardon me. I forgot that it is forbidden to speak in the language of the pagans, here.'

Kanwar Sahib: You probably do not make such mistakes when you are among the gods.

Doctor Sahib: Then sir, penalize me for the sin.

Kanwar Sahib: The penalty is that you must speak to us barbarians in the mother tongue.

Doctor Sahib: You are a raja, you can meet this condition. I am continually in contact with the English language, I will not be able to keep this promise. And also, as you know, this language is the glory of our country today.

Kanwar Sahib: It is honourable gentlemen like you who have given it its present status. When the unruly soldiers of Persia and Kabul, and the Hindu traders mingled, the Urdu language came into existence. If the educated people of our different provinces had been forced to depend on their own language for mutual communication, a national language would have emerged by now. As long as the educated among us remain under the spell of English, no national language can come into being. But this is a time-consuming task, and who has the time? Especially since we now have access to a complete language like English, people have simply fallen for it. Now everybody is saying that English is our lingua franca, and that the whole idea of giving this status to an Indian language should be dropped. I don't understand why people consider it so prestigious to speak and write in English. I too know the language and have lived in England for two years. Moreover I can speak and write it better than many of its lovers can. Yet I hate it as though it were the secondhand garment of some Englishman.

Padam Singh did not take part in these discussions. As soon as he got a chance, he went up to Bithal Das and showed him Shanta's letter. Bithal Das asked him, 'What do you want to do?'

Padam Singh: I can't think. Ever since I got this letter I feel as though I were floating on the river.

Bithal Das: Something will have to be done.

Padam Singh: Yes, but what?

Bithal Das: Bring Shanta.

Padam Singh: The family will disown me.

Bithal Das: Let them. That's where duty lies and it can't be ignored.

Padam Singh: You are right, but I do not have the nerve to annoy my brother.

Bithal Das: Don't keep her in your house, keep her in the Widows' ashram. That shouldn't be difficult.

Padam Singh: Yes, that is a good idea. My mind seems to stop working in times of trouble.

Bithal Das: But you will have to go there yourself.

Padam Singh: Why? Could you go?

Bithal Das: Why would Oma Nath agree to send her with me?

Padam Singh: What objection could he have?

Bithal Das: You talk like a child sometimes! Even if Shanta is not his daughter, he is still her guardian. Why would he send her off with a stranger?

Padam Singh: Don't be angry with me. I am a bit confused at the moment. But if I go, this matter will become serious. If my brother hears of it, he will kill me. I cannot forget how he pounced on me at Jenwa.

Bithal Das: All right you needn't come, I'll go. But you must write a letter to Oma Nath, I trust you'll have no hesitation in doing that?

Padam Singh: You'll think I am a fool, but I don't have the courage to do that. Think of a way I can extricate myself if something happens. My brother should not have the opportunity to blame me.

Bithal Das said with irritation, 'My imagination is not so resourceful. You call yourself a man! On the one hand you make such impassioned speeches, full of profound feeling, and on the other, these absurd fears!'

Padam Singh said with embarrassment, 'Say what you like but you will have to take the responsibility for this operation.'

Bithal Das: All right, but will you send a telegram or is that also too much to ask?

Padam Singh leapt at this. He said, 'Yes, I'll send the wire. I knew you'd find a way. Now if the subject comes up I'll say I never sent the telegram, somebody else must have done it under my name.'

But the next moment he changed his mind. He felt ashamed of his own timidity, and he thought, 'My brother is not so narrowminded that he should be angry with me for doing the decent thing. And even if he does lose his temper I should not care.'

Bithal Das: Send the telegram today, then.

Padam Singh: Should I accompany you?

Bithal Das: That would be most appropriate. It would tie up everything nicely.

Padam Singh: Then we will both go.

Bithal Das: So, when?

Padam Singh: I'll send the telegram today, and we can take the train the day after tomorrow.

Bithal Das: That is settled, then?

Padam Singh: Yes, definitely.

Bithal Das looked approvingly at his simple-hearted friend, and then the two men went in to listen to the music of the jaltarang* which was echoing pleasantly all around them.

– 19 –

When, in order to improve our health, we go for a change of air, we take good care to eat on time, rest, exercise. We do not forget our goal which is to improve our health. Saman who had gone to the Widows' Ashram in order to mend her spiritual health, never lost sight of her objective. She served her widowed sisters diligently, and when she had the time she read religious books. She also went for sacred immersion in the holy Ganges, every day. These activities gave peace to her wounded heart.

Bithal Das had kept from her the news about events in Amola. But when it was decided to bring Shanta to the ashram he thought it fit to prepare her for her sister's arrival. When he returned from Kanwar Sahib's function, he told her the whole story.

A profound silence had settled on the ashram. It was late at night but Saman was unable to sleep. For the first time she could see clearly the consequences of her wrongdoing, and she was miserable. She was like the patient who, on recovering from the effects of chloroform, sees the gaping wound of his operation and faints from the horror of it. She could see her parents and her sister sitting before her. Her mother's head was bowed with shame and misery. Her father was staring at her with furious, bloodshot eyes. Shanta was a picture of dejection as she looked sometimes at the sky and sometimes at the ground. Saman got up from her bed like a wounded bird and struck her head against the wall. At that moment she saw herself as a demon. The blow to her head was a hard one and she staggered and fell.

After a moment when she came to, she found that her head was bleeding. She opened the door of her room quietly. It was pitch dark outside. She ran towards the gate but found it locked. The old watchman was asleep by the gate. Saman went up to him quietly and began to look for the key on his bed. But the watchman woke up and began to shout, 'thief, thief', whereupon Saman ran headlong to her room and locked the door. Then she burst into sobs, cursing herself for ruining her family by trying to satisfy her greed. 'I have murdered my father, I laid the knife on my sister's neck! Alas, how can I show her this blackened face! How miserable Father must have been to hear about my misdeeds!' This last thought was more painful to her than all the others. If instead of telling her father about her, Madan Singh had had her crushed in a mill, stamped upon by elephants, tossed into a blazing fire, thrown to the dogs, she would not have demurred. When she was leaving her house it had not occurred to her that she would have to live in the red-light district; she had left without thinking anything through. In her state of despair she had forgotten that she had a father and a sister. The long period of separation from them had blurred their memory and she considered herself alone and unaccountable to anyone in this world. She had thought, 'I am living in a world where I have neither friends or relatives. Whatever I do here will never be known. Nobody will ridicule or blame me.' But now she found herself tied once more in relationships whose existence she had forgotten. This re-emergence of close ties lighted the torch of her sense of honour.

Saman spent the rest of the night in spiritual torture. At four in the morning, as soon as the gate was opened, she left for her bath in the Ganges. Since she frequently went alone, the watchman did not question her.

When she arrived at the river she began to look about her. She had not come for the ritual bath that day, she had come to drown herself. She felt no fear, no apprehension, or anxiety. Shanta was to come to the ashram next day and it was simpler to hide her face for ever in the river's lap, than to face her sister.

Suddenly she saw a man coming towards her. The darkness had not lifted completely but Saman could tell that the man was a sadhu. There was a ring on her finger which she thought she would give to the sadhu, but when he came nearer Saman hid her face with shame, contempt, and horror. The man was Gujanand.

Gujanand fell at her feet and said in a trembling voice, 'Saman, forgive me.'

Saman moved back a step. The scene of that night when her life was pushed to its ruin took shape before her mind's eye. The wound reopened, and she felt like heaping abuse and humiliation upon him. She wanted to say, 'You are my father's murderer! You ruined my life!' But the look of contrition and humility on Gujanand's face, his fakir's garb, and her own present inclination to forgive, together melted her, her eyes filled with tears, and she said, 'It was not your fault. It is all the fruit of my own wickedness.'

Gujanand: No, Saman don't say that. It is the fruit of my ignorance and stupidity. I had hoped to atone for it, but seeing its fatal consequences I have come to think that my repentance will not have much effect. Alas, these eyes saw Pandit Krishan Chandr drown in the Ganges.

Saman said agitatedly, 'Were you there?'

Gujanand: Yes, I was there. It was night and I was on my way to Amola. I met him on the way. Seeing him go in the direction of the river late at night, I became suspicious, and persuaded him to come with me to where I was staying. I tried to soothe him, and thinking that I had been successful, I fell asleep. After a little while, when I woke up, I found him not there. I ran towards the river and it was then that I heard him call, but even before I could decide on the direction from which I had heard his voice, the wicked waves had done their work. That pure soul departed for heaven under my very eyes. Only then did I realize fully the enormity of my sin. Only God knows how I will be punished for it.

Gujanand's spiritual despair acted on Saman in the same way as soap affects dirt. It lifted the dusty cloud from her heart and banished the thoughts that she had concealed. She said, 'God has enlightened you. You may yet be forgiven and know peace. But what will become of me? Both, this world and the next are lost to me. Alas, the greed for a life of luxury has destroyed me. It is useless to hide this from you now, that your poverty, and more than that, your harshness filled me with frustration at my circumstances. At that time, I was at my most vulnerable, and your love, sympathy, and gentleness could have soothed the blister in my heart, but instead, you crushed it with brutality. I was wounded to the core. When I remember your cruel behaviour, the spark in my heart begins to blaze and I curse you from the core of my being. These are my last moments, in another instant this sinful body will have drowned in the Ganges. Therefore I pray now to God that he should forgive you so that you can live to atone for my, and your own, sins.'

Gujanand said in a worried tone, 'Saman, if one could atone for one's sins by putting an end to one's life I would have done so long ago.'

'At least one's troubles would cease.' Saman replied.

Gujanand: Yes, your troubles would cease, but not of those who are hurt by your troubles. Your parents are free of the bonds of life, yet their spirits are hovering around you. They are still affected by your happiness and your troubles. Remember that by taking your own life you will be heaping more sin on them, and if you lead a good life they will be at peace. Remember also what effect your death will have on Shanta, a girl who still does not know the ways of the world. Who can she turn to but you? Oma Nath, as you know, will not be able to get her married. There is compassion in him, but more than compassion there is greed. Sooner or later he is bound to get rid of her; and then who can she turn to?

Saman could sense true concern in Gujanand's words. Therefore she looked at him humbly and said, 'That is just why I have decided to drown myself. It seems easier for me to drown myself rather than face Shanta. She has written a letter to Padam Singh telling him that Oma Nath is arranging her marriage to somebody else. It is not acceptable to her.'

Gujanand: She is a noble girl.

Saman: What could poor Padam Singh do? He decided to bring her to the ashram and keep her there. If his brother agrees to take her, well and good, but if he doesn't, who knows how long the poor afflicted girl will have to remain there. She will go there tomorrow. The thought of facing her, the shame of it, kills me. What will I do when I see condemnation in her eyes? And if out of hatred she decides not to embrace me I will surely poison myself. It is better to die than to face such humiliation!

Gujanand looked at Saman with veneration. He felt that under similar circumstances he would have suffered exactly the feelings that she was going through. He said, 'Saman, what you are saying is true, but for Shanta's sake you will have to bear all your suffering. Nobody can help her as you can. So far you have been living for yourself alone, now is the time to begin living for others.'

Saying this Gujanand went back the way he had come. Saman stood by the Ganges for a long time, pondering over his words. She then bathed in the river and went towards the ashram, like a defeated soldier with his head bowed, going home from the battleground.

— 20 —

Although Shanta had written to Padam Singh, she did not have much hope of getting a reply from him. Three days had passed and her despair deepened. 'If his answer is not positive, Oma Nath will definitely marry me off to another.' She trembled at the thought. Many times in each day she went to the temples to pray. Never forgetting Sadan for a single moment, she would talk to his picture, saying, 'Sadan, why don't you take me? Do you fear for your reputation? Alas, my life is so worthless! You are abandoning me, casting me into the fire, just because I happen to be Saman's sister! Is this justice? If I could find you, I would grasp you and not let you get away. You are not a stone, you would melt if you saw my tears. If only you could see me, you would give way. Yes you would certainly give way. Your large heart cannot be empty of compassion. What can I do? How can I show you the state of my heart?'

On the fourth day, early in the morning, Padam Singh's telegram arrived. Immediately Shanta was overcome with apprehensions, her ardour for Sadan waning. Anxiety for her future life filled her heart with doubts.

But Oma Nath was ecstatic. He sent for musicians, invited the whole village and made arrangements to receive them. The villagers were amazed—what kind of a sendoff was this? There was no proper ceremony and the bride was being sent off. They thought Oma Nath must be up to his tricks. The whole business was a fraud!

But when the time came, Oma Nath went to the railway station and brought back his guests, amidst the full blare of the musicians' instruments. The new arrivals consisted of just three men, Padam Singh, Bithal Das, and a servant.

The next evening the ceremony for the bride's send off was to take place. When by late evening not a single woman from the village had arrived, Oma Nath was furious. He cursed and threatened, but to no avail. One's kith and kin cannot tolerate high-handedness indefinitely. They do not miss such an opportunity to repay haughtiness and pride.

The palanquin bearers arrived with the palanquin. Janhvi and Shanta embraced each other, sobbing. Shanta's heart was brimming with love and the pain of leaving behind her past. For the time being at least, she had forgotten all the misery she had borne in that household. The thought of a final parting was breaking her heart.

Janhvi too felt waves of compassion in her heart. 'This poor orphan girl has suffered at our hands', this thought made her tears flow, and both women were overwhelmed by fine and sincere feelings.

When Oma Nath came in, Shanta fell at his feet, sobbing, and saying, 'You are my father, don't forget me. Give my two sisters jewellery and clothes, call them home for Holi, but remember me sometimes, just a few words from you is all I ask for.'

Oma Nath consoled her, 'You are my daughter, like the other two. May God keep you happy always.' Saying this he too wept.

It was late evening, and the cow, Munni, came home from her grazing. Shanta put her arms round her neck and cried. She had looked after the cow for three or four years, running after her with the hay, and tying black thread* woven with seashells around her neck. Munni licked her hands, the pain of parting was clearly showing in her eyes.

Janhvi brought Shanta to the palanquin and helped her in. The bearers picked up the palanquin, and Shanta felt as though she was being carried into deep water.

The village women, who were standing at their doorsteps, wept to see the palanquin leave.

Oma Nath accompanied his guests and his niece to the railway station. When it was time to leave, he took off his turban and placed it at Padam Singh's feet.* Thereupon Padam Singh embraced him.

When the train began to move Padam Singh said to Bithal Das, 'Now begins the most difficult part of this drama.'

Bithal Das: I don't get your meaning.

Padam Singh: Are you going to take Shanta to the ashram without telling her anything? She needs to be prepared.

Bithal Das: You are quite right. So, should I go and tell her?

Padam Singh: Think first what you are going to say. She believes she is being taken to her husband's home. This hope is sustaining her in her unhappiness. How miserable she'll be when she knows the truth! I wish I had divulged all this earlier.

Bithal Das: So what's wrong in telling her now? The train stops at Mirzapur. I'll go in and explain everything.

Padam Singh: I made a grave mistake.

Bithal Das: If regretting your mistake makes you feel better then regret it to your heart's content.

Padam Singh: Give me a pencil. I'll explain it all to her in a letter.

Bithal Das: No, why not send a telegram. What a fuss you make over a simple matter!

Padam Singh: I can't help it; it is a serious situation. I have an idea though: the train has a long stop at Mughalserai, I could go to her compartment then, and tell her everything.

Bithal Das: That's quite ingenious. It is said that one should think before one acts. You come up with good ideas, but only after wasting a lot of time. Why didn't you think of this before?

Shanta was travelling in the women's compartment. Two Christian ladies were also travelling with her. When they saw Shanta these two ladies began to talk about her to each other, in English:

'She is going to her husband's home, it appears.'

'She is weeping as though somebody were forcing her to go there.'

'She has probably never set eyes on her husband, how can she love him? She is scared, poor girl.'

'It is such a vulgar custom they have! They send their girls off to total strangers.'

'It is a residue of that barbaric age when they used to take girls away, forcefully.'

One of the two ladies then turned to Shanta and asked, 'Are you going to your husband's home?'

Shanta nodded slowly, and said, 'Yes.'

'You are so pretty, is your husband equally good-looking?'

Shanta replied with maturity, 'One does not look for comeliness in a husband.'

'What if he is black and ugly?'

'He would still be my lord and master.' Shanta replied proudly.

'Suppose there are two men before you, one is handsome, and the other is not. Who would you prefer?'

'Whichever my parents choose', Shanta answered imperturbably.

Shanta felt that the two women were ridiculing Indian marriage practices, so she thought it was necessary to counter their implicit criticism. After a while she asked them, 'I have heard that you choose your husbands yourselves?'

'Yes, we have total freedom in this matter.'

'You consider yourselves wiser than your parents?'

'How can our parents know whether we will love the husband they have chosen?'

'So you think that love is of prime importance in a marriage?'

'Yes, of course. Marriage is a bond of love.'

'We consider marriage a bond of faith. Our love follows our faith.'

At nine o'clock the train reached Mughalserai. Bithal Das helped Shanta out of her compartment, and seated her on a rug which he had spread on the platform. It would be half an hour before the train to Benaras was ready to leave.

Shanta watched hundreds of people with large bundles on their heads, struggling to leave through one door. Hundreds more were pushing and shoving, in their attempt to come in through a narrow door. Meanwhile, through a backdoor, there emerged Englishmen, hand in hand with their women, twirling their sticks, and followed by their dogs. They came in easily, with nobody to stop them or say a rude word to them. In fact, even the employees of the police bowed to them, greeting them respectfully.

Shortly, Pandit Padam Singh came over to Shanta and said to her, 'Shanta, I am your father-in-law.'

Shanta stood up and folding her hands, greeted him with utmost respect.

Padam Singh said to her, 'You must be surprised that we didn't get off at Chinar. The reason is that I have still not talked to my brother about you. When I got your letter I was very worried, and it seemed to me that what needed to be done before anything else, was to bring you away from Amola. I didn't get a chance to talk to my brother about it and so for a few days you will have to stay in Benaras. I have suggested that you should be put up at the ashram where your sister lives these days. Staying with Saman need not cause you any distress. You should banish from your mind the shameful rumours that you have heard about her. She is a noble woman now, and lives a model life. If that weren't so I would never have agreed to have you stay with her. It may take me a month or two to bring my brother round to accepting you. However if you don't like this arrangement tell me so frankly in order that I may find a place more to your taste.'

Padam Singh had said this with much difficulty. He was himself not very convinced of Saman's praiseworthiness. Moreover, Madan Singh's eventual consent was not as likely as he had made it sound. On the whole he had said more than he meant to say, and it made him miserable to think that he had misled the poor, innocent girl.

Shanta fell at his feet, weeping. All she could say was, 'I am your daughter-in-law. Do what you think is best for me.' No other words could have expressed better her helplessness, her humility, and her misery.

After this scene, Shanta felt unburdened of apprehensions about her immediate future. Her feelings were like those of a man who is happy to see his hut set on fire because it releases him, at least for some time, from his terror of the dark.

At eleven o'clock they arrived at the ashram. Bithal Das went in to inform Saman of their arrival, but found that she was lying unconscious with a high fever, while several inmates of the ashram were tending to her. Many other women received Shanta and escorted her while she crossed the courtyard slowly, and entered Saman's room. Standing by her bed she said, 'Sister.'

Saman opened her eyes. Shanta was looking at her with shocked, tearful eyes. She was thinking, 'Is this my beloved sister who used to play with me only a few years ago? Where are those long black tresses, that radiant golden face, those smiling eyes full of allure, the rose-red lips, the daintiness, the elegance? Where has it all vanished? Is this Saman or her corpse?' Shanta's heart brimmed with love and sorrow. She motioned the other women to move away, and then she hugged Saman, weeping, and said to her, 'How are you, sister? Look at me, I am Shanta.'

Saman opened her eyes, and seeing Shanta, said with a wild expression, 'Is that you, Shanta? Move away and don't touch me. I bring bad luck, I am blighted. You are pure, don't come near me, run away from me.' Saying this Saman lost consciousness again. Shanta sat by her bedside all night, fanning her.

– 21 –

Shanta had been in the ashram for more than a month but Padam Singh had still not told anybody in his family about her. Sometimes he thought he should write to his brother, sometimes he considered going to meet him, at other times he felt he ought to send Bithal Das to speak to him, but he could not make up his mind about any of these approaches.

On the other hand, his friends were pressing him to present the proposal for the expulsion of the courtesans before the Board. They were certain of its success and were worried that the delay might raise some new obstacle in its path. Padam Singh had managed to fend off these people so far, but with the beginning of May, Bithal Das and Romesh Dutt pestered him so much that he was forced to formally announce his intention of presenting his proposal to the Board, and was assigned the day and time for the presentation.

As the day drew closer, Padam Singh's confusion of mind grew. He felt that nothing could be achieved by the acceptance of the proposal unless it had the sympathy and assistance of all the prominent citizens. Only then could it be implemented. For this reason he wanted to gain the support of Haji Hashim. This gentleman had so much influence in the city that even the prostitutes could not go against his wishes. Finally Haji Hashim gave in. He had become convinced of Padam Singh's sincerity.

The day came for the proposal to be put before the Board. The courtyard of the Municipal Board was crowded with people. The opposing side had come out in full force and armed to the teeth with their arguments, to attack the proposal. What would be the plight of the Board was something that could only be guessed at.

The proceedings began. All the members were seated. Dr Shyama Charan had postponed his visit to the hill station. Munshi Abul Wafa had not slept all night, he was seen to be going in and out restlessly. His enthusiasm and concentration had exceeded the bounds of moderation.

Padam Singh presented his proposal and explained it in carefully considered words. He had divided it into three parts:

1. The prostitutes would be evacuated from the main areas of the city to a remote habitation.
2. They would be forbidden from entering parks and public gardens.
3. Dance functions would be heavily taxed. Moreover, under no condition would such functions be held in public places.

Professor Romesh Dutt supported the suggestions.

Syed Shafqat Ali said, 'I agree with this proposal, but I will not support it without an appropriate amendment. In my opinion, the following words should be added to the first part: 'except those who, within nine months, either marry or learn a skill that can earn them a livelihood.'

Kanwar Sahib said, 'I agree entirely with the amendment. We have no right to consider these people contemptible. We, who take bribes day and night, practise usury, suck the blood of the poor, strangle the helpless, have no right to despise any section of the people. We ourselves are the most contemptible. We are the worst sinners—we who consider ourselves civilized, cultivated, and important. It is because of our cultivated compatriots that the red-light district is prospering. It is these people from among us who have trained the birds of pleasure, we ourselves have created this species. In a nation where the tyrannical landlord, the corrupt official, the blood-sucking usurer, and the selfish friend are respected, why should the prostitutes not prosper? What should tainted wealth be spent on, if not on tainted pleasure? The day bribery, usury, and illegitimate gains are proscribed, the red-light area will be deserted, and those birds of pleasure will fly away—and not a day before that. Without the amendment, the proposal is like a wound which has no palliative. I cannot accept it.'

Pandit Parbhakar Rao said, 'I don't understand what the amendment has to do with the original resolution. You can present it as a separate proposal. Whatever you want to do to reform those people is very commendable, but it can be done equally well outside the city. In fact it would be easier to do it outside.

Munshi Abul Wafa said, 'I am in complete agreement with the amendment.'

Munshi Abdul Latif added, 'The proposal is not acceptable without the amendment.'

Dina Nath Tivari also pressed for its acceptance.

Syed Tegh Ali had this to say: The amendment could kill the purpose of the original proposal. You are closing the main door of a house only to open a backdoor. It is not possible that women who have lived a comfortable life will readily agree to earn their livelihood through hard work. The result will be that they will take advantage of this amendment. Some will display a Singer sewing machine to give evidence of their toil, some will exhibit a machine for making socks. Others will open a betel leaf shop, and some will stack fruit, and all this will be a front for business as usual. Not only this, but fraudulent marriages will become the order of the day and, concealed behind them, the nefarious activities will thrive even more than before. To accept this amendment is tantamount to expressing ignorance of human nature.

Hakim Shuhrat Khan said, 'I agree with Mr Tegh Ali's views. First of all these loathsome women should be expelled from the city. After that, if they are seen to be living a moral life, earning an honest livelihood, they should be allowed to return to the city on a probationary basis—after all the city is open to all. I am absolutely certain that the amendment will destroy the whole purpose of the proposal.'

Mr Shakir Beg said, 'Expediency may be commendable in national affairs but in matters of morality it will never do. It can only veil immorality, not eradicate it.'

Pandit Padam Singh was listening silently to the views of both sides. He was impressed by Syed Shafqat Ali's reasoning while Kanwar Sahib's realistic approach strengthened this impression. It seemed to him that it would be wholly cruel to turn the women out of the city. The objections the others were making to the amendment did not convince him at all. After a period of perplexity he said, 'Since the purpose of this proposal is not to harm those women, but to reform them, I have no hesitation in accepting the amendment.'

The chairman of the meeting put the amendment to the vote. There were eight in favour of and eight against it. Seth Balbahadar Das was for it and Dr Shyama Charan did not speak for either side; and the amendment was passed.

After this, views were sought on the first part of the original proposal. Nine were in favour of and eight against it, so the part was passed.

Professor Romesh Dutt, Mr Rustum Bhai and Pandit Parbhakar Rao considered the passing of the amendment as a personal defeat, and they looked at Padam Singh as though he had betrayed them. On the other hand Munshi Abul Wafa and his friends looked extremely pleased with what they considered was their victory. Their delight was anathema to Parbhakar Rao and his supporters.

Views were sought on the second part of the proposal. This time Parbhakar Rao and his supporters opposed it, because they wanted to punish Padam Singh for his deviation. When this part was rejected, Munshi Abul Wafa and his coterie were ecstatic.

The third part of the proposal was discussed. Kanwar Sahib supported it, and Hakim Shuhrat Khan, Syed Shafqat Ali, Sharif Hasan, and Shakir Beg also concurred. But Parbhakar Rao and his friends opposed this last part too. After the endorsement of the amendment they felt that further effort on the matter was useless. These three gentlemen belonged to the school of thought which believes in the principle of 'all or nothing.' So this part of the proposal was also rejected.

Sometime in the night the meeting concluded. Those who had been afraid of losing came out laughing, while the faces of those who had been certain of victory expressed sadness.

While they were leaving, Kanwar Sahib asked Rustum Bhai, 'Why did you do this?'

Rustum Bhai replied sardonically, 'We did what you did. You punctured the water vessel, we broke it.'

Everybody had left. The darkness became more profound. The watchman and the gardener too closed the gates and left. But Padam Singh remained sitting on the grass, a look of anxiety and sorrow on his face.

– 22 –

Padam Singh was not ready to believe that he had been wrong in accepting the amendment. He had not imagined that his friends would oppose him so fiercely on a minor point. He did not regret the rejection of the two parts of his proposal as much as the fact that he had been blamed for it. To him this showed the narrow-mindedness and shortsightedness of his supporters. He still considered the amendment to be incidental, and the apprehensions he had heard about its misuse did not convince him at all. It was beginning to dawn on him that expectations from the implementation of the proposal were not likely to be fulfilled, given the contemporary social set-up. Sometimes he even regretted having taken up the issue. If the onus for the ineffectiveness of the proposal fell on the amendment, he would feel much happier. But there was not much chance of that. He was bound to be blamed for it.

He thought, 'My opponents will ridicule me, they will criticize me for my course of action, and I will be the sole victim of this notoriety. I have no friend. I had hoped for justice from Bithal Das. I had thought that he would bring round the friends who are displeased with me, but instead, he is blaming me.' Of all his erstwhile supporters, only Kanwar Sahib continued to console him.

Padam Singh stayed away from the courts for a full month. He would sit, alone and sorrowful, in a state of anxiety. Opposition from his friends made him wonder what hope there was for the nation if such intelligent, alert people could deviate from their own declared principles. Even if he had been wrong and stupid in accepting the amendment, how could his stupidity force them to behave in this perverse manner?

During this period of anguish, it was for the first time that Padam Singh realized what solace a woman could provide. Sobhdra was the only person who could understand his state of mind, and she considered the amendment more important than he himself did. In his opinion Sobhdra was incapable of understanding and weighing profound issues. But although he considered her opinions to be merely an echo of his own, he drew strength from her support.

Even before a month was over, Pandit Parbhakar Rao began to write a series of articles on the subject. These articles were laced with such innuendoes concerning Padam Singh that he was agonized. There was one article in which his acceptance of the amendment was linked to his past. In another article his course of action was reviewed in the following words: The modern 'servants of the people' may forget the people but they never forget themselves. In the guise of working for the good of the people they stalk their own quarry. Young men may fall into the pit but Haji Sahib of Benaras must be kept happy.

Padam Singh was as surprised as he was infuriated by this slander. It was his first experience of the limits to which people will let their resentment take them. The man who was persecuting him was supposed to be a paragon of decency and civilized behaviour, yet he was so petty minded. Worst of all, nobody had the courage to challenge him.

That evening Padam Singh was sitting at his table, trying to write a rejoinder to the article. He was sitting listlessly, not knowing what to write, when Sobhdra came and said to him, 'Why are you sitting here in this heat? Come and sit outside.'

Padam Singh: Parbhakar Rao has abused me roundly today. I am trying to write a rejoinder.

Sobhdra: Why is he victimizing you?

Saying this, Sobhdra picked up the newspaper and read the whole article in five minutes.

Padam Singh: What do you make of it?

Sobhdra: This is not an article, it is just a string of abuse. I thought that only women indulged in such squabbles, but it seems that we are being left behind by men. This gentleman must be learned too?

Padam Singh: It is difficult to gauge the depths of his knowledge. There is nothing under the sun on which he cannot give his views.

Sobhdra: And yet he can write this!

Padam Singh: I am writing a very strong rejoinder. It will remind him who he is dealing with!

Sobhdra: But how can you respond to abuse?

Padam Singh: With abuse?

Sobhdra: No. The response to abuse can only be silence. Even the ignorant can counter abuse with abuse. What then is the difference between you and them?

Padam Singh looked at Sobhdra with appreciation. Her advice had found its mark. Sometimes we receive good guidance from people whom in our arrogance we think are unintelligent. He said, 'So shall I not say anything?'

Sobhdra: That's my view. Let him say what he likes. He is bound to be ashamed of it one day. That would be his punishment.

Padam Singh: He is never embarrassed. Such people are immune to feelings of that sort. If I see him now he will talk to me as though nothing had happened. But by the evening he'd be working on the invective again.

Sobhdra: Is it his job to censure others?

Padam Singh: No, that is not what his job requires, but people like him take delight in thrilling their readers with such fabrications. Such malice make the sales of the publication soar dramatically. The public loves such sensational stuff, so the editors ignore their higher duty in order to indulge their readers' taste for scandal. In fact, some have no hesitation in saying that it is their duty to keep the public happy, that they must please those whose patronage provides them their livelihood.

Sobhdra: Then these people are merely the slaves of wealth. They deserve pity more than wrath.

Padam Singh left the table. He decided against writing a rejoinder. He had never thought Sobhdra so wise, but he realized now that in spite of his superior education he did not possess a nature as generous as hers, that notwithstanding her ignorance, her mind was more alert than his. He became aware that even a childless woman can be a source of peace and tranquillity for her husband. Sobhdra had awakened a new feeling of love in him which cleared away the resentment of many years. He looked at her with gratitude and affection. The look was not lost on Sobhdra, whose heart filled instantly with happiness.

– 23 –

When Sadan came home after seeing Saman, he was like the poor man who has been robbed of his life's savings.

He thought, 'Why didn't Saman speak to me? Why didn't she look at me? Does she find me so repulsive? No, perhaps she is ashamed of her past and would like to forget me. Or perhaps she has heard of my marriage and considers me unjust and cruel.'

Once again he felt a great urge to see her. The next day he again went towards the bathing place that was nearest to the Widows' Ashram, but changed his mind and turned back halfway. He couldn't think of what to say if Shanta were mentioned. At the same time he remembered the sadhu Gujanand's admonition.

Sadan had begun to think of his duty towards Shanta—months of listening to speeches on social issues was bound to leave its trace. In his heart he had begun to acknowledge that he and his family had been unjust to Shanta. But he had not yet developed the strength of character that enables a man to be guided by his conscience despite the prospect of ignominy.

He had become fond of reading and study, and had formed a dislike for the marketplace and the red-light district. In the meetings of the Arya Samaj* he had heard many speeches which dwelt on the importance of morality and civilized behaviour. He had been told that education alone was not a proof of moral correctness, nor was it more important than it. The effect of all this was to give him a taste for publications on moral reform, which he devoured. He never forgot for a moment the suggestions for controlling and vanquishing personal desire that abounded in such literature.

Sadan had been present when the proposal for the expulsion of prostitutes had been presented in the Municipal Board. He considered the amendment injurious to the proposal and thought his uncle in the wrong for accepting it. But when Parbhakar Rao began to shower criticism on Padam Singh, he began to support him impulsively. He wrote and posted several articles to Parbhakar Rao and for a long time entertained the hope that they would be published. He was certain that his articles would cause an upheaval, perhaps start a revolution. Every day, as soon as the newspaper arrived, he would scan it for his articles but instead of them he would find the same censorious and heartbreaking writing, day after day. It stung him to read it, but it was the last article which proved too much for his patience. The editor needed to be taught a lesson, he thought, no matter what the consequences. 'Had he been objective he would definitely have printed my articles. The fact that he has not done so shows that he is not interested in getting to the bottom of the matter but merely in entertaining his readers with lies and exaggeration.' However he did not divulge his thoughts and plans to anybody.

In the evening Sadan went to the office of the *Jagat*, armed with a thick rod. The office was closed but Parbhakar Rao was sitting in his part of the office, writing something. Sadan went in boldly and stood before him. Parbhakar Rao looked up to find a well built young man standing before him, with a staff. 'Who are you?' he asked the stranger angrily.

Sadan: I live close by. All I've come to ask is why you have been abusing Padam Singh for so long.

Parbhakar Rao: So it's you who sent me those two or three articles? Thank you for them, and sit down, I have wanted to meet you. Your articles are factual and well reasoned, and I would have printed them long ago except that it is against our policy to print anonymous articles written by people we cannot identify.

Sadan gave his name. His wrath had subsided.

Parbhakar Rao: You are an admirer of Padam Singh?

Sadan: I am his nephew.

Parbhakar Rao: I see. How is he? He hasn't been here in a long while.

Sadan: He is well so far, but I don't know what will become of him if your articles continue to pour abuse on him. Since you used to be his well wisher and supporter, why have you become so mistrustful of him?

Parbhakar Rao: Mistrustful! Why do you say that? I am not at all mistrustful of him. Perhaps you are not familiar with the duties of us editors. We consider it an obligation to make the public privy to our views. It is against our practice to conceal our feelings. We are neither the friends nor enemies of anybody. In matters of national importance we never pardon the mistakes of anybody because this would make those mistakes even more pernicious. Padam Singh is a close friend of mine and I respect him greatly. I differed from him on matters of principle only, but recently I found some evidence which indicates that he had his own secret reasons for accepting the amendment. I don't mind telling you this: many months ago he had a prostitute called Saman Bai admitted secretly into the Widows' Ashram. For the last one month her younger sister too has been lodged there. Much as I want the report to be baseless, I will be printing it, if for no other reason than that it should be contradicted.

Sadan: Who has told you all this?

Parbhakar Rao: I am not at liberty to tell you that. But please tell Padam Singh that if the allegation is a false one he should let me know. I know that before his proposal was presented to the Board he used to visit Haji Hashim every day. In such circumstances you will realize to what extent I can be convinced of the innocence of his intentions.

Sadan was no longer angry. Parbhakar Rao's vindication of his stand had been effective. At heart he was convinced of his sincerity, so after a while he took his leave. At this point his greatest concern was whether Shanta had in fact been brought to the ashram.

At dinner that night he tried hard to broach the subject, but his courage failed him. An anxiety kept him awake all night. Why was Shanta in the ashram? Why did Padam Singh send for her? Had Oma Nath refused to keep her in his house? Such questions kept coming up in his mind. Early the next morning he went to the ashram's bathing area, hoping to meet Saman there and ask her for an explanation. He had been sitting for a while when he saw Saman come towards him. Behind her was another woman, whose face was veiled.

Saman stopped when she saw Sadan. For many days now she had been looking for a chance to speak to him. Even though she had decided earlier never to address him, it was necessary to break her resolve for the sake of Shanta. She said to Sadan shyly, 'Sadan Singh, what good fortune it is to see you today! You have stopped coming here, is everything all right?'

Sadan replied awkwardly, 'Yes, everything is all right.'

Saman: You have lost weight. Have you been ill?

Sadan: No, I am very well. Death refuses to touch me.

Saman: Don't say such ill-omened things. It is I who should be courting death, I who am the cause of all this. Why are you standing? Sit down, I need to talk to you today. Pardon me, I will be calling you Brother now. I am now your sister-in-law. If you don't like something that I say, don't let it upset you. You must know my circumstances. It was your uncle who rescued me from my sinful life and now I live in the ashram, weeping at my past, for which I will ever weep. My unfortunate sister joined me about a month ago. She was not welcome in Oma Nath's home. May God bless Padam Singh; he went to Amola himself and fetched her. But having brought her here, even he has forgotten her. Now I ask you, is this justice that one sister pay for the other's crime? You know well the whole situation: by reason of my own bad luck, circumstances, or sins in a previous incarnation, I left the path of faith and for that I had to be punished and was punished. But what crime has this poor girl committed for which you have

abandoned her? You have to answer my question, and do not put the
blame on your elders; only cowards do that. Now tell me honestly,
was this not callous? And how did you allow such cruelty? Did it not
grieve you at all that you were ruining the life of a helpless girl ?

Had Shanta not been there Sadan might have had the courage to
say what he felt and own up to his cruelty. But he was not ready to
make this confession in Shanta's presence though at the same time he
was ashamed to justify his actions in the name of family honour. He
did not wish to say anything that might grieve Shanta, nor did he want
to utter words that would raise her hopes. The quick look he had cast
on Shanta had put him in a dilemma. His state of mind was of a boy
who looks longingly at a sweetmeat brought by a guest but dares not
eat it for fear of what his mother might say. He said to Saman, 'Bai, I
don't know what to say. How do I tell you that whatever took place
was done at the bidding of my elders, without pinning all the blame on
them and appearing irreproachable myself? I too was afraid of the
scandal. Even you will agree that if one lives in this world one must
adopt its ways. I agree that injustice was done, though we were only
its unwilling instruments, the whole force of the society we live in
was upon us.

Saman: Brother, you are an educated man, I can't win with you.
But cruelty is cruelty, whether it is one man who commits it or whether
the whole race is responsible for it. It should never be perpetrated out
of fear of anybody. Shanta is here so I don't want to give away her
secret, but I have to say this that even if you find more wealth, beauty,
and connections elsewhere, you will never get the kind of love that
she could give you. It is the love that has made her refuse to find
solace and comfort in another marriage, it has brought her here. No
matter what you do she will always be yours and never belong to
anyone else.

Sadan could see tears flow from Shanta's eyes. His heart was full
of empathy for her. He said impulsively, 'I don't know what to do.
God alone knows how grieved I feel.'

Saman: You are a man, you can do everything.
Sadan: I am ready to do whatever you bid me.
Saman: Is that a promise?
Sadan: Only I know how I feel. What can I say?
Saman: One cannot trust what men say.

Saman smiled when she said this, and Sadan replied with embarrassment, 'I would tear out my heart and show it to you if I could!'

Saman: All right, then take Shanta's hand, and face the river. Say to her, 'You are my wife and I am your husband. I will always protect and support you.'

Sadan's moral courage gave up and he began to look sheepish. It seemed to him that the river was about to swallow him. He looked at the sky like a drowning man and feeling keenly his own brazenness, said hesitatingly, 'Saman, give me a chance to think.'

Saman replied sternly, 'Yes, think well. There is no hurry. I don't want to put you in a moral dilemma.' Then she turned to Shanta and said, 'See, your husband is standing before you. I have said what I could but it has not melted his heart. You will be losing him for ever, so if your love is sincere and there is any strength in it, hold him and make him pledge his word.' Saying this, Saman went in the direction of the river. Shanta followed her slowly; her love had been vanquished by pride. She had hardened towards the same man for whose sake she had decided to endure all affliction, to whom she had given her heart. She gave no thought to his problems, it did not occur to her that he was not yet his own master. At this moment if she had spoken to him with humility her hopes would certainly have been realized—Sadan was not hard-hearted. But she found pride more appropriate than supplication.

Sadan stood there for a few moments, and then with a heavy heart took the direction of his house.

– 24 –

Sadan felt as though he had committed a grave sin. He recalled again and again what he had said, and came to the conclusion that he had behaved ruthlessly. The wounds of love had made him extremely susceptible.

'Why do I fear the world so much?' he was thinking, 'What has the world given me?

'Because I dread scandal, should I deprive myself of this blessing which is the reward for good deeds done during God knows how many past incarnations? Shall I deny myself this treasure which can make me forget all the blessings of the world? What do I care if my kith and kin disown me for doing what is right! Fear of scandal is good as long as it stops us from doing wrong, but if comes in the way of doing right, then it is cowardice to succumb to it. If we initiate court proceedings against an innocent man the world doesn't blame us, in fact it helps us by providing us with witnesses and lawyers. If we swindle somebody of his money or property the world does not punish us, or punishes us only slightly, but for something like this which is not even remotely reprehensible, the world doesn't spare us, it denounces and maligns us. So, for fear of such a world and its evil tongue shall I abandon her, leave her to drown in midstream? No, the world may say what it likes, but I cannot bring myself to inflict such injustice.

'It is true that it is my duty to obey my parents. They brought me into this world, raised me. I can drink poison at their bidding, jump into the flames, but their insistence will not make me strike an innocent woman, least of all a woman who I have given my word to protect and support. My parents will be angry with me. They may even disown me, consider me dead. But after some time their grief and anger will lose its edge, and they will forget me. Life will heal their wound.

'Ah, how harsh I am! This beautiful girl who is exhilarating as the morning light, and delightful as the twilight, stood before me with bent head, like a helpless supplicant, and I did not soften! I should have laid my head on her feet and begged her forgiveness. I should have brought water from the Ganges and washed her feet with it, but instead, I stood there like a statue talking pompously of family honour!

'Alas! how grieved her sensitive nature must be by my excuses and evasions. Her look of detachment is proof of that. She must have thought me dry, proud, treacherous, for she did not even glance at me. Yes, I deserve such contempt.'

Such feelings of remorse troubled Sadan for a long time. Finally he decided that he ought to build a shack for himself and become independent. He knew that his parents would reject him, would be deaf to his pleadings. His uncle on the other hand would welcome him, but then he did not want to be the cause of a rift between him and his father. There was no option left but to stand on his own feet.

Every day he would leave his uncle's house with the intention of clearing the misunderstanding with Shanta, but then he would think, 'Where can I take her? What do I have to offer?'

This sad predicament was no less obsessive than the pangs of love. He was always thinking of how to resolve his problem, but could not come up with a solution. He would visit offices and make repeated

trips to factories and workshops, but would return without
accomplishing anything. So far he had lived a carefree life, and had an
attitude of proud detachment. He had never learned to solicit work
although he knew that the world required men to bow their heads in
supplication; it obliged those only who know how to beg, who are
hard working and diligent, who have learned to control their anger
like politicians, who treat reproofs as though they were favours, who
can lap up degradation as though it were milk, and who have got rid
of their dignity, crushing it under their feet. He did not know that the
qualities which are associated with angels were of no value in this
world. He was honest, sincere, and straightforward. But he did not
know that whatever their moral value, the world regarded his qualities
as worthless compared to connections and influence.

Sadan now regretted that he had never learned a skill by which he
could earn his livelihood. His efforts to find work took up a month
and proved futile in the end. His repeated disappointments engendered
frustration, and he began to feel angry at his parents, his uncle, the
world, and himself. He began to resent all show of wealth and
influence.

Only a short while ago he himself used to go out in a phaeton
regularly, but now whenever he saw someone riding in a phaeton, he
felt his blood boil. If he saw a fashionable man walking on the road,
he would keep pace with him, walking unreasonably close to him and
waiting to pick a fight with him if he dared to open his mouth in
protest. Often, he would pay no heed to the screaming of coachmen,
refusing to let a carriage pass him on the road. 'These people are all
dressed up in fine clothes and are out to have a good time. Am I their
slave that I should get pushed to the side of the road, in order to let
them pass?' he would think.

Since his family owned a reasonable amount of property, Sadan
had never been troubled with thoughts of earning a livelihood. But
recently, since he had come face to face with the problem, he realized
that he did not possess the know-how to solve it. Although he knew
no English, he was quite proficient in Urdu and Hindi. He considered
the sophisticated section of society with its indifference to the native
languages to be the enemy of the nation and country. Ever since his
articles had been published in *Jagat* he had become contemptuous of
Anglicized Indians. 'They are all the slaves of their own interests.
They learn English just to exploit the poor and make money. One is a
lawyer, another goes round in a hat and collar—they all suck the

blood of the nation, but claim to be its leaders! They are all the slaves of fashion, whom education has taught merely to ape the British. They have no empathy, no faith, no love for their mother tongue. They have no sense of moral uprightness, no self respect, and no national pride. What kind of human beings are they?' Such thoughts used to cross his mind, but when he joined the search for livelihood he realized that he had been unnecessarily resentful of such people. In fact they were to be pitied.

Then his mind would again assess the situation: 'I know better Hindi than most people; the qualities of my character may not be pure but they are better than those of many; my ideas may not be lofty, but they are not base either. Yet all doors of opportunity are closed to me. I may be able to get a peon's job, or at best I could be a constable. That is all I am worth! We suffer such injustice; no matter how strong our character, how wise we are, how alert of mind, if we don't know the English language, such perfections count for nothing. Who can be more unfortunate than we who put up with this injustice in silence. We should simply give up the thought of finding employment.'

Sadan was like a man wandering in a forest, fretting and fuming in the dead of night.

Strolling by the river in this state of anxiety and hopelessness, he reached a place on the riverbank where a number of boats were anchored. A few small boats were flitting about on the surface of the river, from some of which he could hear the sweet sounds of music. Boatmen were carrying goods out of several of the others. Sadan went and sat in one of them. Soon, the captivating poetry of the endless breadth of water and the thought provoking solitude of the evening took possession of his being, and he began to think, 'What a pleasant life this is. I wish I could live in such a place, rocked gently by the waves and singing happy songs! I could have a shack by the river where Shanta would be waiting for me. And sometimes we could both ride a boat like this one, on the river!' His imagination painted such an attractive picture of a simple and contented life that he was carried away with enthusiasm. Everything around him seemed to be soaked in harmony and poetry. He got up and called out to a boatman, and asked him, 'Is there a boat for sale here?'

The boatman was drinking coconut water from its shell. He got up at once and showed Sadan several boats. Sadan chose a boat and began to negotiate the price. Many other boatmen gathered around them and finally a sum of three hundred rupees was agreed on. It was also settled that the seller would be employed by Sadan to row the boat.

When Sadan left for home he was in seventh heaven. He could not sleep all night. The picture of the boats with their sails spread, coming towards him from the horizon, kept him awake. His imagination had created a charming hut by the river, lush with climbers and surrounding greenery. In this hut then appeared the lovely picture of Shanta. The hut turned into a palace with a beautiful garden, and in the groves of the garden Sadan was strolling with Shanta. On one side was the song of the river, on the other the chirping of the birds. Those whom we love, we always see in the same attitude—the attitude that has become engraved in our memory. When he thought of her, Sadan always saw Shanta in the same simple sari, with head bent, standing by the Ganges. This picture stayed in his eyes.

It seemed to Sadan that there could be nothing but gain in the line of work he was thinking about. The likelihood of loss did not even enter his head. The strangest thing was that he had given no thought to how he would find the money to pay for the boat.

It was only the next morning that he began to worry about procuring the money. Who should he ask? Who was likely to give it to him? On what pretext should he ask for it? Should he ask his uncle, but then Padam Singh was himself constrained for money those days, since he had not been to the courts for months. And of course it was useless to ask his father. But what would the boatman think if he didn't take the money to him soon?

Sadan began to pace on the roof. His castle in the air that he had just created, began to topple. The hopes of youth are like fire in a haystack—it lights up promptly, and dies down with the same rapidity. Or, they are like the dignity of a poor man—it rises in the morning but has collapsed by the evening.

Suddenly Sadan had an idea, and he began to laugh. 'What a fool am,' he thought, 'My mohan mala is lying in my trunk. It must be worth a little over three hundred rupees. I could sell that. If somebody questions me I'll just tell them that I have sold it. But nobody is likely to remember.' He took the necklace out of his trunk and began to wonder how he could sell it without being observed, since to be selling jewellery was not much different from selling one's dignity. He was worrying about this when Jitan came into the room to sweep the floor. Noticing Sadan's anxious expression he said, 'You look depressed and your eyes are swollen. Did you have a sleepless night?'

Sadan: Yes, I could not sleep. There is something that is bothering me.

Jitan: Tell me what it is.

Sadan: If I tell you it will be all over the house.

Jitan: I have served you all my life. If I didn't know how to keep a secret, I wouldn't have lasted here so long. You can rely on me.

Sadan spoke with much reluctance and hesitation, 'I have a mohan mala. Sell it for me. I need the money.'

Jitan: That's not difficult. You needn't worry about it. But why don't you ask the mistress for the money? She won't say no. If you ask the master you may not get it, but it's the mistress who runs this house, the master has nothing to do with it.

Sadan: I don't want to ask anyone in the family.

Jitan looked at the necklace, weighed it in his hands, and promised to sell it the same evening and bring the proceeds to Sadan. But Jitan did not go to the market, instead he hurried to his quarters. Once inside his room, he shut the doors and moving his cot to one side, began to dig the ground underneath. Soon an earthenware pot appeared. This vessel contained his savings of a lifetime. The thrift, frugality, dishonesty, brokerage, embezzlement, that he had practised all his life, were sealed in that receptacle, in the form of currency. Perhaps that was the reason why the money was stained and marked. How small were the wages of long, sinful years! How little is the worth of sin!

Jitan counted the money and made piles of twenty rupees each. There were seventeen piles in all. He then weighed the necklace on a scale, against money. It weighed a little more than fifteen rupees. The price of gold had gone up in the market, but he decided to value it at twenty-five rupees a tola.* He then made thirteen piles of twenty five rupees each; fifteen rupees were left over. His total savings were thirty-five rupees short of the value of the necklace. He thought, 'What a bargain! I can't let it go. I'll tell him the weight of his necklace came to no more than thirteen rupees. That will save me another fifteen rupees. And so, Your Highness, pretty Necklace, go rest in your coop.'

The vessel went underground again. The rewards of sin became even smaller.

Jitan was in ecstasy. He had 'made' fifty rupees in a matter of minutes. He had never had such good luck in his life. He thought he must have set eyes on a good man when he woke up that morning. In the wreckage of honesty, as in a blinded eye, light is unknown.

At ten o'clock, Jitan brought three hundred and twenty-five rupees and put them in Sadan's hands. To Sadan the money seemed like a gift from heaven. He offered five rupees to Jitan, saying, 'Buy yourself some tobacco.' Jitan made a face and said, 'I have everything, thanks to your generosity. What will I do with this?'

Sadan: Don't refuse. It is my pleasure to give you this.

Jitan: No, young master. If I had been so greedy, I would have been a rich man today. May God grant you much more.

Saying this Jitan went away, and Sadan was left with the impression that here was a good and upright man. He decided that at some future date he would reward him.

In the evening, Sadan's boat was floating on the waters of the Ganges as clouds float on the sky. Instead of happiness, the expression on his face showed anxiety for the future. He was like the student who having passed his exams, is overwhelmed with depression because he knows that the dam which had protected him so far from the floods of the world is now broken. Sadan was thinking, 'My boat is on the river but its course will not always be smooth.' He had realized that the water was deep, the wind strong, and the voyage of life perhaps not as pleasant as he had thought. If the waves sang sweetly, they could also roar in a fearsome voice. The wind which generally lulled the waves could also throw them up with a frightful ferocity.

– 25 –

Parbhakar Rao's fury which had partly abated when he read Sadan's articles, disappeared altogether when, at Sadan's insistence, Padam Singh wrote to him, chronicling Saman's history.

It was three months since the proposal for the expulsion of the prostitutes had been accepted. But the concerns of Tegh Ali and others regarding the amendment had been proved wrong. No new shops were opened over the houses in the red-light district, nor did the prostitutes show any inclination for entering into matrimony. Instead, many of them left their salons, and moved to other quarters for fear of being expelled forcefully. Parbhakar Rao was reassured.

Padam Singh had taken part in a movement born out of hatred, and the proposal had emerged from it as its outcome. But a close analysis of the issue had replaced repulsion with humanity and sympathy—

feelings that had forced him to approve of the amendment. His line of thought was: 'These poor women are the slaves of passion. Lust has blinded them. In their situation they need our compassion and sympathy. If they are treated with harshness, they will lose whatever capacity for reform they might still have, and these souls whom we could reclaim with love and good counsel will forever go outside our sphere of influence. We who are ourselves the slaves of our baser instincts have no right to punish them. The harm that they are doing themselves is so great that it does not need to be compounded by us.'

Thoughts precede actions. Padam Singh mastered his reluctance and stepped into the world of action. The same man who had stood before Saman could now be seen sitting in the parlours of the red-light area. He was no longer afraid of what people might say, he had no dread of ridicule or contempt. His conscience had awakened, and he was inspired by a sincere desire to serve. Raw fruit will not leave the branch even when hit by stones; on the other hand, when it ripens, it will fall to the ground on its own.

Bithal Das did not share Padam Singh's views on the matter. He had no faith in the reformation of corrupt souls. Syed Shafqat Ali who was the author of the amendment, evaded Padam Singh, and Kanwar Sahib had no time to spare from his engagement with music. Only Swami Gujanand helped Padam Singh, sacrificing himself completely in the service of others.

– 26 –

A month had passed, but Sadan had not told anybody at home of his new occupation. He would get up early every morning and on the pretext of bathing in the Ganges, would go and tend to his boat. He would come home at ten o'clock, eat, and then go back again. He returned home only after nightfall. His boat was decorated better than any of the other boats. He had spread a rug in it and placed a few stools over it. Tourists from the city preferred to hire his boat for rides on the river. Sadan did not speak to them about the fare; all the negotiations were carried out by his employee, the boatman Jheengar. He himself would go and sit under a tree or on another boat. He often told himself that there was no shame in earning a livelihood; at least he did not have to wait upon anyone, and no one could reprimand him, however he conducted his business. Yet whenever he saw a well-to-do person come in the direction of his boat, he would beat a retreat and his eyes would lower in shame. After all he was the son of a landlord, the nephew of a lawyer, and to have become a boatman—how low he had fallen!

Such foolish scruples cost him dear. He would agree to take only half of the fare to which he was justly entitled. His wares were good, and it only needed an elegant vendor for his business to become a success. Style makes an impact, no matter what the trade. If a well-dressed barber cuts your hair you feel compelled to pay him more. A well-groomed coachman is a candidate for a generous tip. Sadan realized this, yet was made helpless by his inhibitions. Nevertheless his earnings amounted to almost two rupees a day and the time drew near when he could build and occupy his hut by the river. He had become financially independent and was able to stand on his own feet. The thought excited him to the extent that he frequently spent the whole night in dreaming about his hopes for the future.

Meanwhile, the Municipal Board had decided to build houses for the prostitutes, outside the city limits. Lala Bhagat Ram was given the contract for the job. However, not finding a suitable site for a lime kiln and for preparing bricks, he had to take his work to the other side of the river. When the material was ready and he needed a boat to bring it back in, he met Sadan by the river. Sadan showed him his boat, Bhagat Ram approved of it, and Jheengar negotiated the fare. It was decided that the boat would make two trips a day to fetch the material. Bhagat Ram paid the earnest money, and left.

Money is addictive. Sadan was no longer the spendthrift that he used to be. He was preoccupied by the urge to make money, and anxious to complete the job he had undertaken. He watched every penny. And as he was avidly looking forward to earning more money and building his shack, he went off to the river even before dawn. He woke up Jheengar and made him unmoor the boat. By dawn they had crossed the river. On the way back Sadan took the oars, and for fun, began to row the boat. Soon an increase in the speed of the boat became perceptible, and he was encouraged to double his efforts. Rowed by a powerful pair of hands the speed of the boat increased twofold. At first, Jheengar had been smiling, but when he saw how fast the boat was moving he was impressed. He realized that the master was not soft and helpless, and therefore he himself was not as indispensable as he had thought. He began to view Sadan with more respect.

That day the boat made two trips. The next day it made only one trip because Sadan was late in coming. On the third day they made three trips—the third trip was completed at nine o'clock in the night. Sadan was drenched in perspiration and so exhausted that going home became an effort. Two months of such hard work brought him a tidy sum of money and he was able to employ two more boatmen.

Sadan was now the boss of several boatmen. His shack was ready. In it were a divan, two beds, two lamps, and a few utensils. It had a sitting room, a bedroom, and a kitchen. Outside was a terrace paved with bricks and surrounded by plants in flowerpots. Creepers were planted in two troughs and they climbed up the walls of the shack and covered them. The terrace had become a meeting place for the boatmen, who frequently sat there to smoke tobacco. Sadan had corresponded with the authorities to relieve them from the oppression of forced labour, and this achievement together with the bold steps taken to accomplish it, had greatly impressed the boatmen. Moreover, Sadan often gave them interest-free loans from his savings. He wanted to buy a bicycle, and also a launch for the entertainment of tourists. He had already placed an order for a harmonium—Sadan was preparing to welcome the being who would turn his hut into a haven that the gods might envy.

Certainly, he was now able to bear the expenses of a household, but he was not bold enough to bring Shanta to his home without the consent of his uncle. Whenever he sat down to a meal with Padam Singh he told himself that he would resolve the matter that very day. But at the crucial moment his courage would desert him, leaving him speechless.

Although Sadan had never told Padam Singh of his incursion into the boating business, he had heard of it from Bhagat Ram. He was immensely pleased by his nephew's industry and would have liked to increase his fleet by a couple of boats, but since Sadan himself had not mentioned the subject to him, he decided not to bring it up. He had always indulged his nephew, but now he began to accord him some respect. But in Sobhdra's attitude towards him there was a manifest difference—she had begun to love him like a son.

One day, Sadan was sitting in his shack, looking out at the river. For some reason the boat was late in coming. A lamp was burning in front of him, but Sadan who had a newspaper in his hand was too restless to read. He was afraid that his boat had had an accident. He put down the newspaper and went out. The moonlight had spread in a

golden sheet over the sand. The moon's rays falling on the still surface
of the river appeared like a spreading pool of fresh water, emerging
from a dark valley. Several boatmen were sitting on the terrace, talking
together. Suddenly, Sadan saw two women approaching from the
direction of the city. One of them asked a boatman, 'We want to cross
the river, is a boat available?'

Sadan recognized the voice. It was Saman's. He felt a tickling in
his breast, and happiness filled his eyes. He ran to the terrace and said
to Saman, 'Bai, what are you doing here?'

Saman looked intently at him as though she didn't recognize him,
then she said, 'Sadan?' Meanwhile the other woman had pulled her
veil over her face and walked into the shadows, away from the light
cast by the lantern.

The boatmen got up and gathered around, but Sadan told them, 'Go
away and leave us alone. These are women from my family. They will
spend the night here.' Then he turned to Saman and said, 'Is everything
all right? What brings you here at this time of night?'

Saman: Everything is all right. It's just my destiny fulfilling itself.
You've probably not read today's paper. There is something in it that
Parbhakar Rao has written, which has caused a commotion in the
ashram. If we had stayed there another day, the ashram would have
been deserted. So we thought it wise to leave at once. We'll be much
obliged if you can find a boat to take us to the other side. Once there,
we can take a tonga to Mughalserai where there are sure to be trains
bound for Amola. There are none leaving from here at night.'

Sadan: You are home now. Why should you go to Amola? I know
you have been through great trouble but I cannot tell you how happy I
am to see you at this moment. I myself would have come to you but
my work left me no time. For the last three or four months I have been
working as a boatman. This shack is your home. Go in.

Saman went in but Shanta remained standing silently in the
darkness, with her head bowed. Ever since the day when she heard
Sadan utter those heartbreaking words, the girl had spent her days
weeping. Again and again she regretted the pride that had prevented
her from appealing to his compassion. Sadan's face would appear
before her and his words would echo in her ears. What he had said
was indeed harsh, but to Shanta his words still conveyed the flavour
of love and sympathy. She had explained it all to herself, 'It is all the
fruit of my unhappy destiny. It is not Sadan's fault at all. How can he
disobey his parents? It is base of me to expect him to be undutiful.

Alas, I treated my lord with arrogance, humiliated him for my own self-interest!' As the days went by, her anxiety and unhappiness took their toll, and the once-beautiful girl dried up like a river at the peak of summer.

When Saman had gone into the shack, Sadan went out slowly and stood before Shanta. 'Shanta,' he brought out, trembling. He could say no more.

The girl became exhilarated with love. She thought, 'God knows how long I will live and whether I will see him again. For once, let me do what I want to, let me place my forehead on his feet and weep. I will never again get a chance to do this. And if he raises me with his hands to wipe my tears, I will be satisfied, my life will have become worthwhile; and as long as I live I will look back on this unexpected piece of luck with happiness. I had never thought I'd see him again, and if God has granted me this chance, why should I throw it away? If in the waterless desert of life, I find this green and shady tree, why should I deprive myself of a short break in its cool shade?'

Thinking such thoughts Shanta fell, weeping, at Sadan's feet. But her broken heart could no longer sustain the burden of strong emotions. A single gust of wind was enough to crumble the faded flower. Sadan pick her up to embrace her and hold her close to him, but one look at her withered face tore from him a cry of pain. When he had first seen her by the river she had been Beauty's freshly blossomed bud. But now she was no more than a leaf afflicted by autumn—dry and yellow.

Sadan's heart was trembling like the reflection of glittering moonbeams in a river. He picked up her unconscious form with unsteady hands. In the sincerity of his remorse the thought of God entered his head. He said, weeping, 'I have sinned, my God, I have crushed brutally a gentle and delicate heart. But this punishment is too harsh, You are merciful, have mercy on me.'

Holding Shanta to his heart, Sadan went into the shack, and placing her on a bed, he said helplessly, 'Saman, see what's happening to her. I'll run and get a doctor.'

Saman looked at her sister, to find her forehead covered with perspiration. Her eyes had glazed over, her pulse was non existent, and her face was the face of death. Saman began to fan her. The anger that had been piling up inside her over the months while she saw her sister's health and spirits decline, suddenly burst. She gave Sadan a furious look and said, 'This is the result of your cruelty. You have done it. Your brutal hands have bruised this flower, your feet have

crushed this little plant. But now release is coming! Sadan, since the
day you said those harsh words she never smiled, her eyes were never
dry. I had to force food on her. And you inflicted all this punishment
on her merely because she was my sister. And you had rubbed your
nose at my feet and stroked the soles of my feet. You were my slave,
in love with me for years. So, weren't you your parents' dutiful son,
then? Weren't you the scion of a noble family, then? Didn't your
unholy actions of those days sully the name of your high-born family?
Eat leftovers in the dark, but refuse a banquet in daylight! It is all
sheer hypocrisy. You will suffer for it in the presence of God. No
other woman would have looked at you when you said what you did.
But she, poor child, loved you in spite of it, continued to love you.
Go, get me some cold water.'

Sadan had been listening to her, his head bowed like a criminal's.
Saman's reproaches made him feel better; if she had abused him,
perhaps he would have been even more relieved. He considered himself
fully deserving of such rebuke. The more serious our sins the more
our endurance of criticism becomes.

Sadan brought a glass of cold water and handed it to Saman, and
began to fan Shanta himself. When after Saman had sprinkled the
water over her face, Shanta still did not open her eyes, Sadan asked
timidly, 'Shall I fetch the doctor?'

Saman: No, don't worry. She will come round when the coolness
gets to her. The doctor has no cure for this.

Sadan felt better at these words, and said, 'Saman, think what you
like, but I tell you honestly that my soul has known no peace ever
since that ill-fated moment. Many times I thought I should beg
forgiveness, but what recompense could I make? What could I offer? I
had no expectation of sympathy from my family, and as you know I
had no skills, and no independence. It became my sole concern to be
able to earn some money so that I could build a shack for myself. For
months I looked for a job but could not succeed in finding one. Finally
fate brought me to the river bank. I felt like throwing myself into a
boat. Either it would take me to the other side or I would drown. But
the boat turned out to be seaworthy. Finally I built my shack, and then
I thought that if I saved some more money I'd build a house in some
village. Is she better?'

Saman's anger had cooled somewhat. She said, 'I don't think she's
in danger now.'

Sadan felt so happy that had an idol been there he would have banged his head on its feet. He said, 'Saman, you've given me new life. Had something happened to her, I would have died too.'

Saman: Be quiet. Don't say such inauspicious things. God willing she'll recover without any medicine, and the two of you will live happily ever after. You are her medicine, and your love is her life. If she has you she will not ask for more. But if ever you hurt her again she will be like she is now and you will lose her.

Meanwhile Shanta turned on her side and asked for water. Saman held the glass to her lips. She drank a little and lay down again. She was looking around her as though not sure of what she saw. Suddenly she sat up and said to Saman, 'This is my home, isn't it?' Yes, yes, it is. And where is he, my master? Call him, so I can see him. Tell him to come and put out the fire. I have to ask him a question. He will not come? I'll go to him myself then. I will quarrel with him today. No, I will not quarrel, I will only tell him, 'Do not leave me again. Just keep me with you, any way you like, but just keep me. I can't bear this anymore. I know you love me. All right, if you don't love me, at least I love you. Not even that? All right, I don't love you, but I am married to you. No? Well then, I am not married to you. But I will live with you all the same. And if you reject me there will be consequences, yes there will. I did not come into this world just to shed tears. Yes, people will laugh at you, but for my sake you must bear that. Will your parents disown you? No, never. Parents never cast off their sons. And I will bring them back. Will they have no compassion for me...?' Shanta's eyes closed again.

Saman: Let her sleep. She'll feel better when she has slept. It's late now, you should go home. Padam Singh must be worried.

Sadan: I won't go home tonight.

Saman: You must, or they will worry. Shanta has recovered now. See, she is sound asleep. It is the first time in a long while that I have seen her sleep like this.

Sadan refused to go. He reclined on the divan which lay in the verandah, and began to think.

– 27–

Bithal Das was a just man. He went unhesitatingly wherever justice led him. When he saw Padam Singh leave the path of truth, he left his side, and did not go to see him for many months. But when

Parbhakar Rao began to attack the ashram and divulged some things from Saman's past, Bithal Das fell out with him as well. He was left without a friend in the whole city. Experience was teaching him that as the head of an institution of public welfare which was dependent on the generosity and support of others, it was entirely inappropriate for him to associate closely with any one of the inmates. It was evening, and he was sitting alone, thinking about Parbhakar Rao's allegations and how he should respond to them. Much of what he had alleged could not be denied. It was after all, a fact that Saman had been a prostitute. And knowing this, he had admitted her into the ashram. He had also chosen to keep the management committee in the dark regarding Saman's past.

He thought, 'I treated the ashram as my personal property. No matter how lofty my purpose, I had no right to keep the facts secret.'

Bithal Das was still thinking along these lines, and had not decided on a course of action, when the ashram's matron came, and said to him, 'Sir, Anandi, Rajkumari, and Gori are preparing to go home. I tried to dissuade them but they are bent on leaving.'

'Tell them they can go. I am not afraid of them. I can't tell Saman and Shanta to leave.' Bithal Das said with irritation.

The matron left, and Bithal Das went back to his thoughts. 'What do these women think of themselves? Is Saman so contemptible that they can't stay in the same place as she? They claim that the ashram has become disreputable, and if they stay here they will get a bad name too. All right, so it is disreputable. Go then, and nobody will stop you!'

At that moment the postman came, bringing five letters for him. One of the letters read, '...I can't leave my daughter Vidyavati in your ashram. I am coming to take her.' Another gentleman had threatened that if the prostitutes were not expelled from the ashram, he would stop his donations. A third letter said the same thing. Bithal Das did not read the other two letters. The threats did not intimidate him, instead, they irritated him even more. 'These people think that they can make me tremble with their bullying! They don't know that Bithal Das does not care for them. I will never send Saman and Shanta away. Even if that destroys the ashram.' Pride had blotted out Bithal Das's sense of what was right. Firmness and obduracy come from the same source as pride. The difference lies in the practice.

Saman realized that she was the cause of the crisis. She regretted coming to the ashram. She had served the widows with great devotion, but it came to naught. She knew that Bithal Das would never turn her out, so she decided to leave the place quietly, without his knowledge. Three women had already left, three or four were preparing to leave. Several women had written to their families, and now only those women were likely to remain who had nowhere else to go. But even these were disassociating themselves from her. Saman could not bear this humiliation, and so she consulted Shanta.

Shanta was in a dilemma. She did not want to leave the ashram without Padam Singh's consent. But when Saman told her decisively that she would leave whether or not Shanta accompanied her, she was left no option but to go with her. She was like a man wandering in a forest, who accompanies another for no reason but to have company.

Saman asked her, 'What if Padam Singh is annoyed?'

Shanta: I'll explain it all to him in a letter.

Saman: Think about it. You may regret your decision.

Shanta: I ought to stay here, but I can't stay without you. But where will you go?

Saman: I'll leave you at Amola.

Shanta: And you?

Saman: Wherever God leads me. I might go on a pilgrimage.

The sisters talked for a long time. Then they held each other and wept. As soon as it was eight o'clock and Bithal Das went home to dine, they left the ashram stealthily.

The next morning when they were found to be missing, the watchman informed Bithal Das. He ran to Saman's room, only to confirm what the man had said. Both the sisters had left. He was angry with Saman for going, and for taking Shanta with her. How would he explain it to Padam Singh? Suddenly he saw a letter on Saman's bed. He picked it up and began to read. Saman had written the letter to him just before she left. It appeased him somewhat, but he still thought that he had been humiliated on her account. He had decided that he would shame those who had threatened him. But now the opportunity for that was gone. What would people think now? That they had managed to intimidate him? The thought depressed Bithal Das.

Finally he left the room, and went straight to Padam Singh. Padam Singh was stunned by the news. He said, 'What should we do now?'

Bithal Das: She must have reached Amola.

Padam Singh: Possibly.

Bithal Das: Can Saman travel so far on her own, easily?

Padam Singh: Yes, easily.

Bithal Das: Saman couldn't have gone to Amola?

Padam Singh: The two sisters may have drowned themselves in the river.

Bithal Das: Why not send a telegram to find out?

Padam Singh: How can I dare to ask? Since I couldn't look after Shanta, what right have I to make such enquiries? I used to trust you implicitly; had I known that you would prove to be so careless, I would have kept her in my own house.

Bithal Das: You talk as though I sent her away on purpose.

Padam Singh: If you had reassured them they would not have left. You broke the news to me when it was already too late to stop them.

Bithal Das: You want to pin the whole responsibility on me!

Padam Singh: Who else can I pin it on? Aren't you the Administrator of the ashram?

Bithal Das: Shanta lived in the ashram for three months. In all of this period, you never once visited her. If you had done so now and then, she would have been reassured. Since you did not trouble yourself, who could she count on? I acknowledge my own blameworthiness in the matter, but you cannot be exonerated either.

Padam Singh had been irritated with Bithal Das for some time. It was Bithal Das who had persuaded him to help in the reformation of the beauty mart, but when the time came for practical work he had evaded the whole issue. On the other hand, Bithal Das looked at him with suspicion for his recently demonstrated solicitude for the prostitutes. Full of resentment towards one another, each was bent on demolishing the other. Padam Singh would have liked to shake him up, but receiving such a crushing reply, he was silenced. All he could say was, 'Well, I am partly to blame, certainly.'

Bithal Das: No, I am not trying to prove that you are the guilty party. It was all my fault. Since you had entrusted her to me it was natural for you to feel at ease.

Padam Singh: No, in fact it was the consequence of my inertness and timidity. You could hardly have forced them to stay.

Padam Singh had turned the tables by acknowledging his mistake. We can make others bend by bending ourselves, not by stiffening our attitude.

Bithal Das: Perhaps Sadan knows something about it. Please send for him.

Padam Singh: He was away all night. He has built a shack by the river, and he probably spent the night there.

Bithal Das: The two sisters may have gone there. Shall I go and look?

Padam Singh: That's most unlikely. He is not so enlightened. He wants to have nothing to do with them.

Suddenly Sadan entered the room. Padam Singh asked him, 'Where were you all night? We spent the night waiting for you.'

Looking at the floor, Sadan replied, 'I am very sorry. I was forced to stay the night and there was no time to inform you. I have been too embarrassed to tell you that for the past few months I have been in the boating business. I have built a shack by the river and now I want your permission to live in it.'

Padam Singh: Lala Bhagat Singh told me about it but it saddened me that you yourself never mentioned it. If you had done so, I may have been of some help. Anyway, I don't think there is anything wrong with it, in fact I am happy to see you working for a livelihood. However, I will not agree to your starting your own kitchen when after all your home is with us. Would another boat enhance your earnings?'

Sadan: Yes, Uncle, that is what I want. But for that it is essential for me to stay there.

Padam Singh: This is not a fair condition. To live in the same city, yet live separately! No, I will not agree to that, even if it means that you make a loss.'

Sadan: Uncle, please grant me this. I am asking you only because I am forced to.

Padam Singh: What is it that is forcing you to do this? Am I a stranger that you will not tell me? Why don't you speak plainly?

Sadan: I can't give you a bad name by staying in this house. I have decided to do my duty now. So far I had been evading it out of laziness and cowardice, but it is no longer possible for me to ruin a life merely in order to uphold my father's prejudices. Nor do I want to spoil your relationship with him by pinning the responsibility of my act on you. I am your nephew, and whenever I need you I'll come to you; but I'll live apart, and I am sure that you will appreciate my views.

Bithal Das had got to the bottom of the matter. He asked Sadan, 'Did you meet Saman and Shanta yesterday?'

Sadan blushed, and said in a low voice, 'Yes.'

Padam Singh: You could have offered me some help, at least. If you went with me for an hour to the red-light district you would realize that what you call 'impossible' is entirely possible. There are good people and bad, everywhere. The prostitutes are no exception to this rule. You'd be surprised to learn how much their aspirations are governed by love of religion, distaste for the repulsive life that they lead, and longing for an improvement in their lifestyle. I know it surprised me. All they need is some support to cling to, while they pull themselves out of the dark recesses of the underground.

At first they avoided me, but when I explained to them that the proposal was in their interest, that it aimed to keep them out of the reach of the depraved and the lecherous, they began to trust me more. I can't mention their names, but there are rich prostitutes who are ready to make financial contributions. There are many who would like to see their daughters married. However at present there are many more who are enthralled by the pleasures and luxuries of their profession, and they ridicule me. They say, 'Allow us to enjoy ourselves now; we'll turn a new leaf when we are old.' But I am hoping that Gujanand's persuasions will make a difference. It is a pity that there is nobody to help me, while there are plenty who are pleased to mock me. At present we need an orphanage where prostitutes' daughters can live and be educated. But I can't convince anyone.

Bithal Das was listening attentively to Padam Singh. His own experiences coincided with Padam Singh's; and one's own experiences are a convincing yardstick. He said, 'Why didn't you speak to Anarudh Singh about this?'

Padam Singh: What could I get there except verbosity and banter?

Bithal Das fell silent.

– 28 –

Sadan's marriage with Shanta was solemnized in their hut by the river. The hut had been decorated. But there were not many people. Bithal Das, Bhagat Ram, and a few others attended the ceremony.

Padam Singh had left for his ancestral home the same day. He gave a full account of the situation to his brother. Madan Singh was wild with fury. He said, 'I'll cut off his head! Who does he think he is?'

Bhama said, 'I'll go at once and talk to him. He is just a foolish boy. He must have been influenced by that woman, Saman. I am sure I can persuade him.'

But Madan Singh scolded his wife. He said, 'If you attempt to go there I will strangle you. If he wants to jump into the fire, let him do so. He is not an infant. He is being stubborn and I'll make a beggar out of him. He thinks that when I die he will live comfortably on my assets. But he'd better know that this is no ancestral property; I bought it with my own earnings and I'll endow it all to charity. He will not get a penny from it.'

There were rumours and gossip all over the village. Beja Nath became convinced that the world had been emptied of morality. How could moral character survive if people stooped to such baseness.

The boy would never be able to show his face in the village again.

Padam Singh sat with his brother late into the night. But whenever Sadan was mentioned Madan Singh gave him such furious looks that he did not dare to open his mouth. Finally when his brother got up to retire, Padam Singh said with a courage born from despair, 'Bhaiya, Even if Sadan stays away from you he will be known as your son. All his actions, good or bad, will be the actions of his family. Those people who are aware of the facts may know that we were helpless, but the rest of the people will judge us and him as one family. So since we cannot avoid sharing his disgrace why suffer a double loss by losing him as well? In my opinion it would be best to go and talk to him, and if he does not agree...'

'If he doesn't agree, we marry him to that witch' said Madan Singh, ferociously interrupting his brother. Then he continued, 'I am surprised that all this took place under your nose and you knew nothing about it. He bought the boat, built the shack, conspired with the two witches while you sat there, blindly. I had sent him there, trusting in you; how could I know that you would be so oblivious to what was going on? And now that the milk is spilt you come to me with your wise suggestions. Let me be quite frank: even if you have not openly supported him, I suspect you of deliberately turning a blind eye. Of course I have always treated you so shabbily that you had to avenge yourself by doing this to me! Anyway, never mind; write out a deed of gift tomorrow morning and except the few assets that I have inherited, I'll make an endowment of everything I have to charity. If you can't make out the deed here you can do so when you get back. I'll sign it and then it can be registered.'

Saying this Madan Singh went off to sleep. But Padam Singh had received such a murderous blow that he stayed awake all night. The blame which he had tried to avoid, sacrificing even his principles, had

their children, or fashion a toy engine or watch for another. If one of the women falls ill, she visits her and ministers to her. Saman is trying to rebuild her fallen support, and she has been successful enough in this to the point that all those poor people praise her and think highly of her. It is only the members of her own family who do not value her. Love is an absorption, a madness, that makes people selfish. Saman looked after the entire household with the utmost devotion, but Sadan never uttered a word of either gratitude or appreciation. Even Shanta did not esteem her for the hard work that she put in. Both took her for granted, as though she were a servant and was expected to do all the menial work. Sometimes she suffered from headaches and sometimes she ran a fever, but she continued to work all the same. But the two people who were closest to her had become so insensitive that they never noticed the condition she was in. Sometimes when she was alone, she would weep for hours. But there was no one to feel for her, no one to wipe her tears.

Saman was by nature a proud woman. All her life she had been treated with respect. She was the mistress in her husband's house even though her troubles there were manifold. Even in the courtesan's district her prestige was high. In the ashram she was held in high esteem for her selfless service to all.

Saman had a taste for distinction; she savoured renown. Therefore it distressed her extremely to live as a non-entity in this household. She would have been happy, and worked even harder for them, had Sadan showed her some appreciation, if he had sought her advice on some matters now and then, treated her as a person of consequence in his household, and had Shanta sat with her and talked pleasantly to her sometimes, or if she had offered to oil and dress her hair. But how could all this happen, for the newly-weds were still drunk on love. When one takes aim to shoot, one's eyes are fixed on the target. Lovers are in a similar situation.

But it is doubtful if Sadan and Shanta's carelessness towards her was entirely the outcome of their absorption with one another. Sadan avoided her as we shun a leper—in spite of our feelings of compassion towards him we lack the courage to go near him. Shanta mistrusted her, she dreaded her beauty. It was well for all concerned that Sadan himself kept his distance from her. If he had not done so, Shanta would have died of anxiety. Both Sadan and Shanta would have liked her inauspicious presence to vanish for ever from their lives. But afraid of being thought petty, they did not dare to discuss the matter even between themselves. Gradually this reality dawned on Saman.

One day Jitan brought a gift from Padam Singh. He had come many times before, but Saman used to hide herself whenever she saw him approach. But this time he happened to see her. Jitan was one of those who could keep down a rock, but never gossip. On the pretext of smoking tobacco he went and sat with the head boatman, and gave him the scandalous news. 'How do you find her? When her husband turned her out she came to us as a cook. When she was dismissed she went and joined the whores. And now I find her ruling the roost here!'

The head boatmen was stunned. Soon the women began to gossip. From that day onwards, no boatman would drink the water in Sadan's house. Their women stopped coming to Saman. One day Bhagat Ram came to settle the account for loading bricks. Being thirsty, he asked a boatman for a drink of water. The boatman brought the water from the well. Sitting in Sadan's house, to have sent for water from outside was like driving a knife into Sadan's heart. The insult was so great.*

By the end of the second year matters reached a stage where, at the slightest excuse, Sadan was irritated with Saman. And even if he did not actually say anything insulting, his tone was harsh. It became evident to Saman that it would be no longer possible for her to live there. She had thought that she would spend the rest of her life with her sister and brother-in-law, taking up no more than a corner of their house, and looking after their house in return, but even this support was slipping away, and once again she was a crushed being.

But no matter how miserable Saman felt at her own condition, she did not blame Shanta and Sadan. Her study of religious books and an honest appraisal of her circumstances had made her humble and tolerant. She pondered over the question of finding a place to go to, where all would be alien and where she would never find any acquaintances. But no such peaceful prospect presented itself. Besides, her poor, weak heart needed support. The thought of having nobody to turn to made her tremble. She could not face the storms of life, alone and unaided. 'Who will protect me? Who will look after me?' These fears prevented her from leaving her present shelter.

One day Sadan came home and said, 'When will the food be ready? Hurry up, I am going to see Pandit Oma Nath. He is visiting Uncle.'

Shanta asked him, 'Why has he come?'

preoccupation, they attacked him with his own munition. Padam Singh stayed away from the Board's session, and Dr Shyama Charan was in Nainital. Both sub-proposals were passed unopposed.

Houses were being built for the courtesans near Alipur. They had been authorized by the Board. Lala Bhagat was supervising them diligently. Some were of clay and others of baked brick. Some houses had two storeys. There was a small market, and a small clinic, as well as a school. Haji Hashim had started building a mosque and Seth Chaman Lal had commissioned a temple. Dina Nath had started a garden. It was hoped that everything would be completed on schedule, but despite much haste it took a year to complete the project. As soon as it was done, notice was given to the courtesans to leave their quarters in the red-light district and move into the new precincts.

The common impression was that the courtesans would oppose the move strongly, but people were pleasantly surprised to find that the women obeyed the order happily. The red-light area was vacated in a single day and the place that used to be humming with activity day and night, was deserted by that very evening.

Mehboob Jan was an old prostitute. She gave away all her assets to the orphanage. That evening there was a grand gathering of courtesans at her house. Shahzadi, who was one of them, said during a speech, 'My sisters, today a new era in our lives, begins. May God bless our intentions and guide us to the right path. We have lived in shame and disgrace too long. We have spent too much of our lives as Satan's prisoners. We have murdered our souls and our faith for too long, and have spent too much time on lust and worldly luxuries. The ground in this district is blackened by our sins. Today God has had mercy on us and relieved us of our burden, and we should give thanks to Him. There is no doubt that most of our sisters are distressed by our expulsion from this place and that they see a bleak future for themselves. I beg to remind them that God has not taken away anybody's livelihood. You have talents that will be valued by many. But even if, God forbid, we suffer adversity, we must not grieve, for, my sisters, the more we suffer the lighter our burden of sin will become. I pray again to God Almighty to illumine our hearts with His Light, and to guide us to the right path.'

Said Ram Bholi Bai, 'We should be grateful to Pandit Padam Singh. May God bless him.'

Zehra Jan added, 'I want to say to my sisters that in the future they should keep in mind what is permitted and what is forbidden. Music is permitted. We should improve our skill in it. But we should stop being the playthings of the lecherous rich. We have been enslaved by our sins for too long, now we should be free. Did God create us so that we should give away our beauty, our youth, our soul, our faith, and our honour, to lustful men? When a rich youth falls in love with us how delighted we are; our madams are pleased, and it seems to us that we have trapped a golden bird. But in truth it is we who are caught in the trap; no, we have knowingly fallen into the trap. He has bought us with his wealth, and we have lost to him the precious thing that is our virtue. In future we should dismiss from our community all those who we find are inclined to wander from the straight path.'

Sunder Bai said, 'Zehra's suggestion is a good one. I too think that if a man begins to visit us we should first observe what kind of a person he is. If we love him and he is also drawn to us, then we should marry him. But if he comes only out of lust, we should immediately turn him out and not sell our honour so cheap.'

Ram Pyari said, 'Swami* Gujanand gave us a book in which it says that beauty is the reward for good deeds in a previous incarnation. Unfortunately we squander on our present hell even the wealth we inherited from the past. Those of my sisters who like Zehra Jan's suggestion, please raise your hands.'

Twenty or twenty-five women raised their hands.

Ram Pyari spoke again. She said, 'Now, those who don't agree with her suggestion, please put up your hands.'

Not a single hand was raised this time.

Mehboob Jan, an old courtesan said, 'I know you will laugh at my pious thoughts which come after a lifetime of sin, but I'll still say today, seven days before I leave for Haj, that it gives me great pleasure to see your determination. May God help you succeed in your good intentions.'

Some women were whispering among themselves. It was apparent from their expressions that the sentiments expressed by the others did not please them. But they did not dare speak their thoughts aloud. Indeed, base thoughts are subdued when they come face to face with noble feelings.

The function came to an end and all the women left for Alipur, like pilgrims heading for a holy place. Darkness fell on the red-light district. There was neither the sound of drums nor the melodic preludes of fiddlers. The pleasing notes of song had ceased and the flocks of

Padam Singh: Yes, he is well. He comes to see me every few days, and I too go over now and then. There is nothing to worry about.

Madan Singh: Does the brute ever mention us, or does he think we are dead? Has he sworn never to come here? Will he come after we are dead? If that is his intention perhaps we should go away and then he could come here and look after what is his. I hear he is building a house there. If he is going to live there, who will live here?

Padam Singh: It's not true that he is building a house—somebody has lied to you. But he has set up a lime kiln; and I hear that he wants to buy some land on the other side of the river.

Madan Singh: So tell him that first he should come here and set fire to this house. After that he can buy the land or do whatever he pleases.

Padam Singh: What are you saying! He has not come here for fear of displeasing you. If today he knows that you have forgiven him, he'll come running. When he comes to see me he talks about you for hours. If you like he will come here tomorrow.

Madan Singh: No, I will not send for him. Who are we that he should come here. But if he does decide to come here, warn him that he should be ready for me. I can't vouch for my temper; I might beat him up with a stick. The fool! He thinks he can show me displeasure. He was not displeased when he used to drool on the Book just when it was time for my puja, when he urinated near the plate from which I ate. My clothes could never remain clean, thanks to him. I yearned for my clothes to stay unstained, but the moment he saw me wearing clean clothes he would leap at me, all covered in mud as he was. Where was his displeasure then? He thinks he can be displeased now! I'll give him the scolding of a lifetime!

The brothers went into the house. Bhama was sitting there feeding hay to the cow. She stood up as soon as she saw Padam Singh, and said, 'So you have come. It's not far where you live but it doesn't occur to you to visit even once a month, just to find out whether we are still alive. Well, is everything all right?'

Padam Singh: Yes, we are all well, thanks to your prayers. Tell me what you have cooked. If you give me kheer, halwa,* and cream, I can give you a piece of news that will thrill you. Congratulations! You have a grandson.

A flush of pure happiness swept over Bhama's sad face. Her pupils dilated like blossoming flower buds. She said, 'I can set you down in a pitcherful of syrup. Have your fill.'

Madan Singh grimaced and said, 'You've given us bad news. God's justice is all upside down! My son has been taken away, and he is given a son! In short, now they are two! I can never win; I can feel him pull. It is true, God is good to those who have done wrong. Isn't that strange? Now I must appease him, seek reconciliation! How old is the child?'

Padam Singh: He is four days old today. I would have come the first day, but I was busy.

Madan Singh: Never mind. We'll get there on the sixth day.* Let's leave early tomorrow morning.

Bhama was elated. She wanted to be prodigal on this happy occasion. Who to give? What to give? She was in a flurry of excitement. She wanted singing in the house, the shehnai at the door, the neighbouring women to be invited. She wished to see the whole village intoxicated with joy. It seemed to her that something extraordinary had happened. She felt as though in a world full of childless women, she alone had a son and a grandson.

A workman came and told her, 'There is a sadhu at the door.'

Bhama sent the mendicant enough food to feed four mouths.

After dinner Bhama sat down with her daughters and they sang for half the night.

– 32 –

Like one who steals a jewel in a fit of greed but is ashamed to look at the stolen object when he comes to his senses, Sadan avoided Saman. He despised, and humiliated her.

After a hard day's work when he came home, Sadan felt weary of his profession. Working with the lime especially exhausted him. He used to think that he had been forsaken by his family on account of Saman. She had made him an outcast when he could have been living comfortably at home. He had even begun to feel disinclined to eat the food that she cooked. He wished that somehow he could be rid of her; and this was the very same Sadan who, in the past, had adored her and could have given his life for her enchanting smiles, her provocative looks, and her exquisite gestures. But now Saman had fallen low in his esteem.

For several years Sadan had not turned his attention to reading or writing. Especially, ever since he had bought the kiln, he did not have the time even to read the newspaper. He had begun to think that

reading newspapers was the pastime of those who had nothing better
to do. Nevertheless, he managed to spend time on arranging his hair
and playing the harmonium.

Sometimes when he remembered the past he would think, 'How
blind I was to have fallen head over heels for Saman.' He now prided
himself on his serene self-assurance. He used to see thousands of
women by the riverside but no more did lewd thoughts cross his mind.
Sadan therefore imagined that he had achieved moral stability.

But when the time for Shanta's confinement drew near, and she
began to keep to her room because of the weakness and dejection that
accompanied advanced pregnancy, Sadan realized that he had been
deceived. What he had thought was moral stability was in fact the
satiation of his desires. Now in the evenings when he came home after
work, Shanta's happy face was not there to greet him. She would be
lying depressed, on her bed. Sometimes she had a headache, sometimes
it was her body that ached, and at other times she ran a fever. She
suffered from nausea too, and her lovely face had become pale. She
was anaemic. Sadan was saddened to see her in this condition and for
hours he would sit by her, trying to divert her. But then he would
grow tired and make an excuse to leave. Lust had raised its head again
and he was experiencing a storm of desire. He began to joke with the
boatmen's womenfolk. He would look at them with longing whenever
he went to the riverside. One day he felt such a crisis of desire that he
bent his steps in the direction of the red-light district. He had not
visited this area for many months, it was eight o'clock in the night,
and he was walking with absorption. He would take a few steps, stop
and think, and turn around. But after he had taken a few steps in the
opposite direction he would turn around again. At that moment he was
like the patient who, seeing a tempting dish before him, gets ready to
attack it without giving a thought to the harm it may do him.

But when he arrived at his destination he found the place deserted.
Some paan shops were open and crowds of young men could be seen
there. Sweetmeat and bakeries were closed. No lovely faces appeared
at the parlour windows, nor could the sound of tambourines, drums, or
the sarangi* be heard. It was only then that Sadan remembered that
the courtesans had been sent away. For a moment he was distressed,
but immediately afterward he was elated—as though he had escaped
the hold of a merciless policeman who he had to follow willy nilly,
and from whose grasp he was unable to flee. And as though when he
arrived at the police station he found it closed, with neither policeman,

constable, or anybody else to hold him there. He was embarrassed at his own weakness in giving way to desire, and the pride he had begun to feel about his moral integrity was crushed.

He wanted to go back but then he decided to take a stroll in the area. When he had walked a few paces he saw the house where Saman used to live. He was surprised to hear the sound of music coming from the house. But when he looked up he saw a sign which said, 'Music School.' Sadan climbed the stairs to the upper storey where he saw the room in which he had dallied with Saman for months. He saw the old faces in his mind's eye, and sitting down on a bench he began to listen to the singing.

About thirty-five people were learning to sing or play some instrument. Some were playing the sitar, some the pakhawaj,* and some the sarood.* An old music teacher was correcting the notes of each player. Sadan became so absorbed by the music that he sat listening to it for half an hour. He felt a great desire to enlist himself but refrained from doing so at the thought of the long distance he would have to come, leaving the two women alone in the house.

He was about to leave when the old teacher began to sing, accompanying himself on the sitar: I beg you O Mother, take Bharat into your care.*

On hearing this song a spring of noble feeling burst forth in Sadan's heart. His heart was flooded with waves of feeling urging him towards the service of his motherland. The medium of his heart pulsated with the notes of the lyric. The spiritual countenance of a woman appeared before his mind's eye. The wretched, emaciated, and sad old man was looking at the woman humbly, and supplicating with his two hands outstretched before him, 'Make Bharat your own.'

Sadan imagined himself ministering to destitute farmers, begging the agents of landlords to have pity on them. the farmers had prostrated themselves at his feet out of gratitude and their women were blessing him with prayers. He was in short the bridegroom in this imaginary bridal party.

When Sadan left this place he had made up his mind to serve the people. But when he had walked a short distance he saw a crowd outside Sunder Bai's house. On inquiry he was told that Kanwar Anarudh Singh was organizing a society for the prevention of cruelty to farmers by their landlords. Suddenly, on hearing this, the great sympathy he had felt for the poor farmers evaporated. He was himself a landlord and wanted to help the farmers. But as a member of the

landlord community it displeased him that the landlords should be
forced into making concessions. He thought that perhaps those people
opposed the rights of landlords and would like to take away their
prerogatives, and therefore 'we should be wary, and ready to protect
ourselves' he concluded. Human nature is so averse to domination.
Sadan did not want to stay there any longer. It was nine o' clock and
he returned home.

– 33 –

It is evening and the sky is spread with the colours of twilight. Puffs
of breeze are ruffling the waves. The waves smile and sometimes
they burst into laughter, flashing their pearl-like teeth. Sadan's pretty
hut is nestling amid creepers and festive blossoms. There is a crowd
of boatmen at the door and their women are sitting inside the house,
singing songs of rejoicing. A mud stove has been dug into the ground
in the yard, on which food is cooking in large cauldrons. It is the sixth
day in the life of Sadan's newborn son hence the festivities.

But Sadan looks depressed. He is sitting on the terrace, watching
the Ganges flow. In his heart thoughts are rising like the waves of the
river. 'Will they not come? It is the sixth day today. Had they wanted
to come they would have done so already. If it had occurred to me that
they wouldn't come, I would not have informed even Uncle. To them,
I am dead. They don't want to have anything to do with me. It makes
no difference to them whether I live or die. On such occasions people
visit even their enemies. They consider me worse than an enemy!
They could have come for the formality of it if not for the love—just
to show the world that they've done their duty. At least I would have
felt that I still have people who I can call my own. All right they
needn't come if they don't want to. It doesn't matter. Once I am
through with all this, I'll go there myself. How beautiful the child is!
Such red lips. He looks just like me though he has Shanta's eyes. How
intently he was gazing at me. I can't say about Father, but if Mother
saw him once she would definitely take him in her lap.'

Suddenly Sadan was struck by another thought. 'What would
happen to him if I died?' Who would look after him? He has nobody
in the whole world but me. No, no, Father would surely have pity on
him! Let me see how much money I have in my savings account at the
bank. Not even one thousand. If I saved at least fifty rupees each
month there'd be six hundred rupees in a year. As soon as I have two

thousand rupees, I will start building my house. Two rooms outside, five rooms inside. A portico with arches, two rooms above the entrance—in a house like that, life would be delightful. I'll make the plinth high—five feet at least—that'll give it a majestic look.'

While Sadan was absorbed in his daydreams it was growing dark all around. Suddenly he saw a carriage approach from the road. The two lights of the carriage were shining like the eyes of a cat in the dark. Who could it be? 'It could only be Uncle. Who else do I have in the world?' The carriage drove up and stopped near him, and Madan Singh climbed out. Behind it there was another carriage.

Sobhdra and Bhama got off from the second carriage. Sadan's two sisters were in it too. Jitan was sitting on the coachbox. He got down and shone the lantern before them. Sadan saw his family but he did not run to greet them. The time had passed when he would have rushed to appease them. Now it was his turn to be offended. He left the terrace and went into the house, as though he had not seen them. He thought, 'They think that I am anxious to see them. But since they have not cared for me, I too don't care for them.'

Sadan was peeping from inside the hut to see what the new arrivals would do. Jitan came to the door and called out. Several boatmen ran to and fro. Sadan came out, and greeting his mother from a distance, stood to one side.

Said Madan Singh, 'You stand as though you don't know me. Ignore me if you please but kiss the feet of your mother and earn her blessings.'

'My touch will taint your soul.' Sadan replied, indifferently.

Madan Singh looked at his brother and said, 'Just listen to him. Didn't I tell you that he has forgotten us? What was the point of bringing us here? His parents are standing at his doorstep and he still has no feeling for them.'

Sadan could no longer maintain his indifference. His eyes filled with tears and he fell at his father's feet. Madan Singh wept too. Next, Sadan fell at his mother's feet. She pulled him up and hugged and blessed him. It was a glorious scene of love and forgiveness. The parents' hearts were overflowing with happiness, while waves of good resolutions concerning his future behaviour towards his parents, were rising in the son's heart.

Sincerity of feeling lit up the dark corners of the soul, so that false pride and the fear of scandal scurried away like insects, making room for truth and humanity.

Sadan was drunk with happiness. He was ordering the boatmen around, to show his power and status. One boatman was bringing out the string beds, another was despatched to fetch water from the Ganges. A third was sent speeding to the market. Madan Singh was delighted. He whispered to his brother, 'He's turned out to be quite clever. I had no idea he'd be so well-established.'

They went into the labour room. Shanta kissed their feet. Bhama picked up the baby. He seemed to her like the incarnation of Krishan.* Tears of joy fell from her eyes. After a while she went to Madan Singh and said to him, 'Your daughter-in-law is beautiful. She's like a rose petal. And the boy looks like the incarnation of God Himself.'

Madan Singh replied, 'Had he not had a noble destiny how could he have drawn Madan Singh to him?'

Bhama: Daughter-in-law seems to be very sociable.

Madan Singh: That is why Sadan had forsaken his parents.

Everybody was busy talking of their own concerns. To nobody did it occur to ask where the unfortunate Saman might be.

– 34 –

S aman had gone to the river for evening worship. When she returned she found carriages outside the door of the hut. She recognized Padam Singh and realized that Sadan's father had arrived. Her feet would not carry her any further. They were as though chained. She knew that she would no longer be welcome there, so she stood rooted, thinking, 'Where shall I go?'

For the past month there had been much unpleasantness between the sisters. The same Shanta who used to be the picture of grief and tragedy while she lived at the Widows' Ashram, was now bent upon making her sister miserable. The pain of thwarted love had in those days made her generous and sympathetic towards her sister. But now that she had found her great love, her heart had become hard and dry like that of the nouveau riche. Her great fear was that Sadan might once again fall prey to Saman's charms. Her sister's devotion to religious rites, her abstinence, and her prayerfulness left her cold. She thought it all hypocrisy. Saman would long to oil her hair or wear clean clothes. Shanta knew this, she used to watch her very closely. If Saman needed to say something to Sadan, she would tell Shanta. For her part Shanta never let Saman out of her sight. She would even come and sit in the kitchen on some pretext. She wanted to get rid of

Saman before her delivery because she felt that once imprisoned in
the labour room she would not be able to keep her vigil over her. She
could bear any adversity but could not put up with this heartburn.

But Saman pretended not to see anything and not to hear anything.
Like a man who is drowning, she could not bring herself to relinquish
her support even though it was no more than a straw. But when she
saw that Sadan's parents had arrived she was forced to renounce it.
Circumstances had achieved what the will could not accomplish.
Saman decided to drown herself.

She went quietly to the back of the hut and began to listen for any
mention of herself. She stood listening silently for half an hour. Bhama
and Sobhdra were talking of this and that. Finally she heard Bhama
say, 'Doesn't her sister live here anymore?'

Sobhdra: Of course she does. Where else can she go?

Bhama: If she comes, tell her to sleep outside. Does Sadan eat the
food she cooks?

Shanta spoke from her room, 'No, I used to cook his meals but
nowadays he cooks his own.'

Bhama: But she must be touching the earthernware and the china.
Have the earthenware thrown away, the china can be washed.*

Sobhdra: There is no place to sleep outside.

Bhama: Even if there isn't, I won't let her sleep here. How can one
trust such a woman?

Sobhdra: No, sister. She is no longer what she used to be. She is
very pious now.

Bhama: A whore, turned pious! Even if she became a deity now, I
wouldn't trust her.

Saman could not bear to listen anymore. She felt as though
somebody had pierced her heart with a red hot rod. She turned around
and went away.

It was pitch dark and the road was not clearly discernible. Saman
went stumbling forward. Where was she going? She had no idea. She
was running desperately and without thinking, like a dog that has
tasted the rod. She wanted to steady herself, but could not, till a thorn
pierced her foot. She held her foot and sat down. She had no strength
to walk any more.

Like a man who gains consciousness after a spell of fainting, she
looked sharply around. There was deep silence everywhere, and
profound darkness. Only the jackals were singing their unmusical song.
The thought that she was alone in this wilderness sent shivers down

her back. It was the first time in her life that she had understood what
it was like to be really alone. Yet while she knew her isolation to be
complete she seemed to sense the presence of many other beings
hovering in the atmosphere around her. She shut her eyes in fear.
Loneliness certainly breeds superstition.

Saman began to think, 'How unfortunate I am! Even my sister
doesn't want to see me. How hard I tried to win her and how little I
succeeded. There is the brand of infamy on my forehead and it can
never be washed away. How can I blame her or anybody else? It is all
my own doing. Alas! How my ankle hurts! How can this thorn be
pulled out? The part of it that was broken is still inside my flesh. How
painful it is. No, I cannot blame anyone. I sowed thorns, how can I
expect a yield of fruit? How blind I was to have murdered my soul in
order to please my senses. Yes, I was troubled. I wanted jewellery,
good food. My life had become unbearable. But those conditions too
were a retribution for my deeds in a former incarnation. And are there
not women who are confronted with far more adversity and yet they
guard their chastity? Sitaji was turned out of her home by Ramchandr
and for years she had to face untold hardships in the forest.* Savitri
bore up in the face of tragedies. But not only are there these far-
fetched examples, take Aheeran of Amola; she went through times of
so much trouble. For years her husband did not return from abroad
while she starved at home. Alas, my beauty has ruined me. The pride I
took in it has brought me to this!'

'Oh God, why do you place thorns among flowers? Why do you
give beauty, only to take away restraint from the temperament? I have
observed that most beautiful women are spirited. Perhaps God tests us
in this way, He wants to polish the soul by throwing it into the fire of
beauty. But alas, our self-centredness blinds us, and instead of shining
we burn in the fire!

What can stop this pain? God alone knows what kind of thorn it
was. What if somebody tries to catch me? Who will hear me if I
scream? Why are those leaves rustling? Is it an animal? Somebody is
coming, for sure.'

Saman stood up. She was sturdy of heart and she mastered her fear.

It was late at night. The cool breezes of early spring were blowing.
Saman gathered her saree around her and rested her head on her knees.
She remembered a day in the same season long ago when she had
been sitting at the door in her husband's house, wondering where to

go. On that day she was jolted by desires, but now an inner peace predominated.

Suddenly she dozed off. Swami Gujanand was standing before her wrapped in his deer skin, regarding her with compassion. Saman fell at his feet and said humbly, 'Swami, save me.'

The holy man patted her head and said, 'That is why God has sent me to you. What is it you want? Wealth?'

Saman: No, my lord. I have no greed for wealth.

Swami: Do you want prestige, then?

Saman: No, my lord, not even that.

Swami: Then listen: If you want salvation, there is one route only that is open to you. That route is the service of humanity. If you serve those who are even more helpless and unfortunate than you, their prayers will redeem you.

Saman's eyes opened. She looked around her, certain that she had been awake, and wondering where the swami could have disappeared. Then it seemed to her that he was standing among the trees with a lantern in his hand. She stood up and limped towards him. She had thought that he was about a hundred steps away from her, but even after she had taken four hundred steps, the grove of trees and the swami who was standing in it with his lantern, were as far away as when she had taken the first step.

For a moment Saman thought that she was still asleep. But no, it was not a dream. She called out to him, 'Maharaj, please wait for me. I am coming.'

She heard him say, 'Come. I am waiting.'

Saman began to walk again, but after she had taken two hundred steps, she sat down, exhausted. But when she looked up, she could see the grove and the swami, still at a distance of a hundred steps. She was overtaken with dread and began to tremble. She wanted to scream but her voice failed her.

In a while Saman collected herself and began to think. What was this mystery? Was she witnessing a phantom? It was a hand from the preternatural that was beckoning to her.

Once more she stood up and began to walk. She had arrived near a town, and she saw the swami go into a shack. The grove of trees had disappeared. Saman felt calmer as she thought that this was probably where the swami lived and she looked forward to meeting him and solving the mystery.

At the door of the shack she called out to him, 'Swamiji, it's me, Saman.'

The shack belonged to the swami, but he was asleep, so Saman got no reply.

She peeped into the shack and saw the swami asleep by a lighted fire. The scene astonished her. He had just come in, how could he have fallen asleep so soon? And where was the lantern? She called out to him loudly, 'Swamiji!'

The swami sat up and looking at Saman in astonishment, he enquired, 'Is it you, Saman?'

Saman: Yes, Maharaj, it is me.

Gujanand: I was just dreaming of you.

Saman said in confusion, 'But you just came in.'

Gujanand: No, I didn't. I went to sleep a long while back, and I did not leave the shack at all. In fact I was dreaming of you.

Saman: And I've been following you from the bank of the Ganges. You had a lantern in your hand.

'I think you are mistaken' said Gujanand, smiling.

Saman: If I hadn't followed you how could I have come here? I had no idea where you lived.

Saman then told him the entire episode.

Gujanand: It is likely. Such things happen sometimes.

Saman: Could it have been a god who, in disguise, led me to you?

Gujanand: That is also possible. All that you just told me was what I saw in my dream—that I was counseling you to be of service to others. Saman, you know that you have suffered at my hands. You know what a vile man I was. It makes me miserable to think of my brutalities. You deserved to be treated with dignity but I was cruel to you. That was the main reason for our troubles. When will God make the people here realize the worth of our women? A woman can live cheerfully in adversity, she can live by the sweat of her brow, as long as she is treated with dignity and love in her home. Without these no woman can be happy even if she lives in a palace. Unfortunately, I did not know this in those days. Later, after you left and I began to keep the company of holy men, I was enlightened, and I began to be concerned for my salvation. I had neither knowledge nor discernment so I decided to serve my fellow beings. This was the simplest path I could follow, and I have been following it to this day. This path brought me contentment and I recommend the same path to you. I have seen you in the Widows' Ashram, in Sadan's house; and in each

place you were serving others with dedication. In your heart there is
love and compassion. These are the qualities one must have in order
to follow that path. Your road is beckoning to you, it has the door of
service open for you. Step in and God will help you.'

Saman could see a spiritual light on Gujanand's face, and a deep
devotion was born within her.

Maharaj, you are my guru now. I give myself to you. I had made
this vow to you once before but out of my foolishness I could not
fulfil it. However I never forgot it, and today I make that vow again
with a truer heart.

Gujanand could see an expression of great sincerity on Saman's
face. He became intensely agitated. The feelings that he had been
trying to subdue, for many years, came to the fore. Once again his
mind turned to the allurements of life. His present life began to appear
dry, unattractive, and lonely. He trembled at the temptation and began
to fear that if such thoughts took root in his mind, all his devotions
and the esteem with which he was now held on account of them,
would be wiped out. So he said to Saman, quickly, 'Do you know that
an orphanage has been opened here?'

Saman: Yes, I've heard about it.

Gujanand: Most of the girls there have been entrusted to us by the
courtesans. There are some fifty girls altogether.

Saman: It is all the result of your efforts.

Gujanand: No, it's Padam Singh to whom the credit must go. I am
just his humble servant. We need a responsible and sincere person for
the orphanage and I believe that you are such a woman. I have looked
far and wide but have not been successful in finding someone
dedicated, someone who is likely to look after the girls as though they
were her own daughters. God has blessed you with wisdom and
discernment, you are compassionate and giving, I ask you humbly to
take on this responsibility.

Saman's eyes filled with tears. Her heart was gladdened by the
thought that Swami Gujanand had such a high opinion of her. She had
never imagined that anybody would have such faith in her and that she
would be singled out for such glorious service. She became certain
that it was actually God who was speaking through Gujanand's mouth.
A moment ago, if she had seen a boy lying in the mud she would not
have gone to him. But Gujanand had banished her reluctance by the
trust he reposed in her. We do not dare to disappoint those who believe

in us and often agree to carry a burden that formerly we would have considered unbearable. Trust generates trust.

Saman said humbly, 'I am fortunate that you consider me capable of such service. It is my heart's desire to be of some use, and although it is impossible for me to reach the high standards set by you, I will nevertheless carry out your commands to the best of my ability.'

Saman's head was bowed while she spoke and her tears began to flow. Her aspect gave away what her tongue had left unsaid. It conveyed to him, 'You are kind to trust me so much. How could an unfortunate woman like me have dreamt of being charged with such a noble responsibility? But God willing, you will never regret your choice.'

Gujanand said, 'I expected no less of you.' Saying this, he got up. Dawn was breaking and the melodious song of the cuckoo could be heard. Gujanand set off to bathe in the Ganges.

Saman came out of the shack. She looked around her like one just awakened from sleep. How agreeable the weather was, how peaceful and refreshing! Was the day about to break on her future life too? Would the first light of dawn gleam for her? Would the sun's golden rays shine on her future?

– 35 –

A year passed. Pandit Madan Singh had been bent on going on a pilgrimage. It appeared that he would embark on it as soon as Sadan came home. However, ever since Sadan's arrival he seemed to have forgotten his resolve. Instead, he was now often seen settling his accounts with tenant farmers, his grandson in his lap. Or he would oversee the work on his fields. Worldly greed had him in its grip once again.

On the other hand, Bhama now had more leisure. She had more time to exchange views with the neighbouring women, because Shanta took care of the household duties.

Pandit Padam Singh had given up his practice. He was now the Chairman of the Municipality, a position that ensured him the kind of occupation for which he had a natural bent. Under his charge the city was showing great improvement. Several new roads were constructed within a year, and three new gardens had been laid out. He was planning to build housing for coachmen and other drivers of vehicles, outside the precincts of the city. Many of Padam Singh's former friends had begun to oppose him and many of his detractors had become his friends. But he himself had come to place great trust in Bithal Das. He would have liked to give him a position in the Municipality, but Bithal Das did not agree to this. He did not want to depart from his principle of disinterested service. He thought that as a Municipality official he would be able to do far less for the city than he could by remaining independent. His Widows' Ashram was doing very well and was receiving substantial assistance from the Muncipality. Bithal Das was trying to start a fund for the peasants so that they could obtain loans at a nominal rate of interest, in order to buy seeds and other essentials for farming. Sadan was his right hand man in this enterprise.

Sadan did not like to live in his village. He had left Shanta in his family home and returned to the riverbank where he was expanding his business. He now owned five boats and was earning a profit of hundreds of rupees every month. He was planning to buy a steamer soon.

Swami Gujanand spent most of his time in the villages. He had dedicated himself to helping poor girls. If he needed to come to the city, he never stayed more than a day or two.

It was the month of Katak.* Padam Singh had escorted Sobhhra to the Ganges for her bath. On the way back they were returning via Alipur. Sobhdra was looking out through the carriage window. Who could live in this wilderness, she was wondering, when suddenly she saw a magnificent building. On the signboard that hung on its door was written in bold letters, 'Seva Sadan.'

Sobhdra asked her husband, 'Is this Saman's 'Seva Sadan'?
'Yes it is,' said Padam Singh, regretting that he had taken this route. Sobhdra was bound to want to see the orphanage, and he would have to accompany her. Padam Singh had never visited Seva Sadan. Gujanand had tried several times to rope him into going there, but he had always found an excuse to avoid doing so. He found it impossible to face Saman whose words when she returned the bracelet to him, he could never forget. He was still haunted by the thought that he was responsible for leading astray a noble and virtuous woman.

'Stop the carriage, I want to see the orphanage.' Sobhdra was saying.

'It's late today. You can come another time.'Padam Singh replied.

Sobhdra: I've wanted to come for the past year. Now that I am here let me see it.

Padam Singh: You could have come whenever you liked. Nobody stopped you.

Sobhdra: Well, if I didn't come before, I am here now. Why can't I see the place?

Padam Singh: I have no objection, only it is late. It is almost nine now.

Sobhdra: It won't take very long. We'll be out in ten minutes.

Padam Singh: You are always so stubborn! I've told you I am getting late.

Sobhdra: You can tell the coachman to drive faster on the way back. That way you'll make up for the time lost.

Padam Singh: All right you can go in; and come back as late as you like. I'll hire a carriage to take me back.

Sobhdra: You don't need to go. Just wait here, I won't be long.

But Padam Singh began to dismount saying, 'I am off. Come when you like.'

Sobhdra understood his reluctance. She had read laudatory accounts of 'Seva Sadan' in *Jagat* many times. Pandit Parbhakar Rao was partial to the orphanage, and so Sobhdra had developed a regard for it. She had begun to respect Saman and wanted to see her in her present circumstances. She was amazed that after falling so low, Saman had become so enlightened that newspapers had begun to notice and praise her work.

Sobhdra dismounted from the carriage and entered the building. Seeing her in the verandah, a woman went in and informed Saman of her arrival. Saman came out almost immediately. She was dressed with extreme simplicity. Gone were the graces, the playfulness, the smiling eyes of earlier times, and in their place the astonished Sobhdra beheld flickers of a new strength, and gravity.

Saman greeted Sobhdra humbly, and said with tears in her eyes, 'It is my good fortune that I see you here today.'

Sobhdra was touched, and embracing Saman, she said, 'I have been longing to come, but there was no time.'

Saman: Has Padam Singh come too?

Sobhdra: He was with me but because he was getting late he hired another carriage and went back.

Saman said sadly, 'Unfortunately for me, he doesn't like to come here. I am very sorry that he hates the orphanage which he himself founded—on my account. It was my cherished wish that both of you would visit this place. Half of it has been fulfilled today. The day the other half is satisfied will be a happy day in my life.'

Saman then took Sobhdra on a tour of the orphanage. There were five large rooms in the building. In the first room some twenty-five girls were sitting on the floor, reading. Their teacher shook hands with Sobhdra. Saman introduced them. Sobhdra was astonished to learn that the lady was the wife of Mr Rustum Bhai, and that she gave two hours of her time every day to teaching the girls.

There were the same number of girls in the next room. Their ages were between eight and twelve. Some of them were cutting cloth, others were sewing, and yet others were busy pinching one another. An old tailor was sitting with them. He showed Sobhdra garments that the girls had sewn.

In the third room there were fifteen to twenty small girls. None of these were older than five years. Some of them were playing with dolls, others were looking at pictures that were hung on the walls. Their teacher was Saman herself.

Sobhdra was then given a tour of the garden where the girls had planted shrubs and flowers. Some girls were even then watering rows of cauliflowers and potatoes. They presented a bouquet of flowers to Sobhdra.

In the kitchen, several girls were cooking. Saman showed Sobhdra pickles and various other foodstuff prepared by the girls.

Sobhdra was delighted by the organization and discipline of the place, and the skill and good manners of its young inmates. 'I could never manage such a large institution' she thought, and 'all the girls do look clean and happy.'

Saman said, 'I have taken on this responsibility but I don't have the strength to carry it out independently. I follow the advice of others. If you perceive any shortcomings, do let me know. It will only benefit the orphanage.'

Sobhdra replied with a laugh, 'Don't embarrass me. Whatever I have seen here has astonished me. Even the Widows' Ashram is not so well organized.'

Saman: Do you mean that?

Sobhdra: Yes, honestly, I find it even better than what was said in the glowing reports which described it. But tell me, do the girls' mothers ever come to see them?

Saman: They do, but I don't encourage too much coming and going.

Sobhdra: And who will marry them?

Saman: That is the dilemma. One's duty is to teach them all the household skills, whether their worth will be recognized, I cannot say.

Sobhdra: Does the Barrister's wife love this work?

Saman: Yes. Not only that, you could say she is the spirit behind this enterprise. I merely obey her.

Sobhdra: All I can say is that had I been capable of doing good work, I would have loved to work here.

Saman: Even your coming here today has annoyed Padam Singh. He'll never let you come again.

No, I'll bring him here next Sunday. I'll teach the girls to prepare the paan and eat it, then go to sleep.

Saman replied laughing, 'You'll find that many of the girls are cleverer at that than you are.'

Ten or twelve prettily dressed girls came out to where the two women were chatting. They sang a patriotic song which delighted Sobhdra, and she gave them five rupees as reward.

When Sobhdra was leaving, Saman said to her in a dejected voice, 'I'll wait for you on Sunday.'

Sobhdra: I'll come, definitely.

Saman: How is Shanta?

Sobhdra: I received a letter recently. They are all well. Did Sadan come here?

Saman: No, but he sends two rupees a month as a contribution to the orphanage.

Sobhdra: I will take my leave now.

Saman: You did me a great favour by coming.

Sobhdra: And I was longing to see you. Your hard work and good management, your cultured manners, they have all impressed me more than I can say. You are truly the pride of womanhood.

Saman said with tears in her eyes, 'I will always consider myself your humble servant. I'll never forget what I owe all of you. If you had not saved me I would have sunk. May God keep you happy always.'

* * * *

Explanatory Notes and Glossary

– PART ONE –

p. 1. *darogha* : A police inspector.

p. 2. *abolition of dowry* : A live issue even in the India of today. To propitiate her in-laws, parents often spend much beyond their means on their daughter's dower.

p. 3. *mahant* : An important quasi religious personage who is often a trustee of lands belonging to a temple.

p. 5. *sacred duty* : The duty to marry off his daughters.

p. 6. *stain a betel leaf* : To prepare a betel leaf for eating it is stained with catechu and lime. Betel nut pieces are then placed on it, and often other condiments as well, such as cloves, cardamom, or anise seed.

p. 9. *tilak ceremony* : See tilak.

p. 10. *tilak* : A red mark upon the forehead, between the eyes or the eyebrows, worn as an ornament, or placed on another to anoint, consecrate, or betroth.

p. 12. *Magh* : Eleventh month of the Hindu calendar, corresponding to January.

p. 12. *bathe in the Ganges* : Bathing in the Ganges is often a rite, observed as thanksgiving or to bring good luck. In this case Oma Nath is pleased that he has found a match for his niece and is about to take the first step in the marriage proceedings by obtaining her mother's consent.

p. 12. *Phagun* : Twelfth month of the Hindu calendar, corresponding to February-March.

p. 12. *puri* : Thin, deep-fried, pancake-like bread.

p. 13. *jalaibi* : A popular sweetmeat, made from gram flour in the shape of a coil and deep fried.

p. 13. *batlee* : A savoury snack.

p. 19. Chait : First month of Hindu year. Corresponds to March-April.

p. 19. *Ram Nomi* : Ram's birthday. See Ramlila.

p. 19. *geru* : Red ochre. Sadhus and other pious men dye their garments with geru and smear their bodies with ashes. Bathing in the Ganges is also an act of worship.

p. 25. *tanzeb* : A kind of muslin.

p. 30. *Dasehra* : Hindu festival in which efffigies of Lanka's ruler Ravana are burnt. Also, in Hindu mythology, the day the Ganges was born. According to Hindu belief, whoever bathes in the Ganges on this day will have ten of his sins wiped off.

p. 30. *Muharram* : First month of the Muslim calendar. The anniversary of the massacre of Hussain and his party, who challenged Yazid, falls in this month. (Hussain was Prophet Muhammad's (PBUH) grandson and Yazid, the immoral son of the Caliph, Mawiah, had declared himself his father's successor.) One of the commemorative rituals that are observed during this month is the singing of *marsiyas*—dirges that describe the death of Hussain and his companions at Karbala.

p. 31. *Forget the expensive jewellery you were going to buy me* : Sadan is being sarcastic. She is reminding him that he could never afford to buy her anything of value.

p. 48. *Sadan kissed his feet* : Among the Hindus, a gesture of respect shown to one's elders or to pious religious persons.

p. 48. *bhabi* : Brother's wife.

p. 48. *dharam shala* : A building meant for a religious or charitable purpose.

p. 66. *jumped into the fire* : The custom of killing or burning to death their wives and children when attacked by a powerful enemy is known as *jauhar*. The Rajputs (the Hindu caste of fighters) do this in order to fight unencumbered by fears for their families' safety, when they intend to fight to the end. Rajput women have been known to commit *jauhar* voluntarily.

p. 66. *swing festival* : A Hindu festival in honour of the god Krishna and his mistress Radha, celebrated at new moon in August for about five days and nights.

p. 74. *kanwar* : princeling. To hide his identity and to seem more worth her while to the courtesan, Sadan pretended to Saman that he belonged to the nobility.

p. 79. *dharam sabha* : A religious party.

p. 79. *Ramlila* : Religious play based on the legend of the Hindu god, Rama.

p. 110. *banarsi safa* : A turban woven with motifs of gold thread.

p. 100. *achkan and dhoti* : A formal coat-like garment, and a stretch of cloth passing between the legs and fastened at the waist.

p. 100. *forum for independence* : Part of the many discussions that were held to raise support for the demand for termination of British rule in India.

p. 104. *paandan* : A box with several compartments to hold the various condiments that are used to prepare the betel leaf for eating.

p. 106. *barber* : The barber, who had free access to several homes and knew intimate details of the inmates' lives, was often approached when discreet enquiries had to be made.

– PART TWO –

p. 108. *ved* : Physician who practises the indigenous Hindu system of medicine.

p. 113. *shehnai* : A musical instrument somewhat like an oboe or bagpipes, associated with weddings.

p. 114. *Bhaiya* :   Brother. Also used as an affectionate or informal term to address any male.

p. 115. *Singing abuse at weddings* : An old custom. The bride's and the bridegroom's parties vie with each other in singing wedding songs, many of which are abusive and meant to provoke the other side with their bawdy humour, into a more lively exchange.

p. 116. *monsoon music in spring* : Inopportune. Indian classical music has themes for every season.

p. 117. *sacred thread* : Higher caste Hindus wear this thread which goes under one arm and round the neck.

p. 117. *pigtail* : Hindus wear a pigtail on an otherwise shaven head to distinguish them from the lower castes.

p. 117. *widows to remarry* : Hinduism forbids widows to remarry. It was customary for the widow to follow her husband into the funeral pyre, when he died. However this practice has now been discontinued.

p. 120. *deerskin* : Hindu ascetics use this as a prayer rug.

p. 123. *banter with the women* : This is an extension of the custom that a man may engage his wife's sisters in lighthearted banter.

p. 127. *Imambargah* : A room or building used as a venue by Shia Muslims for their commemorative observances of the martyrdom of Prophet Muhammad's (PBUH) grandson Hussain, during the Islamic month of Muharram.

p. 134. *in this same school* : In the company of courtesans.

p. 134. *Tansen* : Title of a famous singer of the emperor Akbar's court.

p. 135. *Ram, Krishan, Shiva, Shankar* : Deities in the pantheon of Hindu mythology.

p. 138. *khasdan* : A small, decorative dish with a lid in which prepared betel leaf is served.

p. 138. *duvar puja* : Ritual worship performed at the door or entrance of the venue of the wedding, when the bridegroom arrives for the marriage ceremony.

p. 147. *distinctions* : An indication of the growing chasm between Hindus and Muslims as early as the first quarter of the twentieth century when this book was written. Abul Wafa, a Muslim, thinks he will be seen as very liberal for doing a favour to the Hindu inmates of the ashram.

p. 147. *Shakuntala* : An outstanding play by the Hindu writer, Kalidas.

p. 147. *political* : This word had become loaded with suggestions of inter racial strife between Muslims and Hindus.

p. 148. *Bhagvatgita* : Hindu scripture.

p. 148. *Rana Partab Singh* : The heir to the thone of the Sikh kingdom in Punjab, who was carried off to London by the British, where he grew up. Out of his regard for Queen Victoria, he gave her the Koh-i-Noor diamond which was in his possession.

p. 155. *ghazal and qawwali* : Ghazal is a romantic poetic form in Urdu poetry. Set to music, it is one of the most popular genres of vocal music in the subcontinent. Qawwali is usually sufic poetry, set to music in a chant-like mode. It is also very popular. Both ghazal and qawwali are lighter forms of music and associated with Muslim culture, although they are equally popular with the Hindus.

p. 156. *even a paan in Saman's house* : In spite of his love for her, Sadan was afraid of soiling his Hindu upper caste purity if he ate or drank in the house of an unchaste woman. Paan is betel leaf.

p. 161. *urs* : Death anniversary of a Muslim saint, celebrated with quasi religious song and dance, in the sufi tradition.

p. 180. *theosophist* : A believer in theosophy, a philosophy which draws its inspiration from Hinduism and Buddhism. The Theosophical Society was founded by two Russians, H.P. Blavatsky and Colonel Olcott. The theosophists deny a personal God, and believe in the transmigration of souls and the brotherhood of man irrespective of race or religion. They have their own theory of the creation of the universe and believe in a complex system of psychology.

p. 180. *Madam Blavatsky* : See *theosophist.*

p. 180. *Col. Olcott* : See *theosophist.*

p. 180. *Tantric Vidhya* : The whole system of rituals in Hindu worship and beliefs.

p. 180. *the Upanishads* : The philosopical parts of the Hindu scriptures.

p. 180. *Gita* : Sacred songs or poetry.

p. 182. *Brahmin, Khatri, Vesh, Shodar* : The Hindu caste system, in which Brahmin is the highest caste and Shodar is the lowest, untouchable caste.

p. 185. *jaltarang* : A musical instrument consisting of glass bowls partly filled with water and struck with sticks to poduce music.

p. 190. *black thread* : A black thread is supposed to protect the wearer against the evil eye.

p. 190. *turban at Padam Singh's feet* : A gesture made by someone who is asking for forgiveness.

p. 200. *Arya Samaj* : A modern reformist but fanatical Hindu sect.

p. 209. *tola* : The tola is equal to 12 *mashas*, and belongs to a weighing system used mostly to weigh gold and silver. It is equivalent to approximately 180 grains troy. It used to be the same weight as the one rupee coin.

p. 227. *Ramayan* : The great Hindu epic poem, *Ramayana*.

p. 227. *Bhagat Mala* : Hindu religious book.

p. 227. *Mira* :   Princess turned religious ascetic, whose hymns are well-known.

p. 229. *The insult was so great* : In traditional Hindu culture, to pointedly not eat or drink in another's house is tantamount to suggesting that one considers it a tainted household. In this case, the discovery of Saman's presence in his house has brought disgrace to Sadan.

p. 233. *swami* : Prominent figure in Hindu religious order.

p. 236. *kheer, halwa* : Sweetmeats. It is customary to offer sweetmeats to the bearer of good news.

p. 237. *sixth day* : The sixth day of an infant's life is celebrated with rituals and feasting.

p. 238. *sarangi* : An indigenous Indian musical instrument resembling the violin.

p. 239. *pakhwaj* : A kind of drum.

p. 239. *sarood* : Lyre.

p. 239. *Bharat into your care* : Bharat is the Hindu name for India. Here the music teacher who is a poor old man is singing a prayer to the 'merciful goddess', begging her to 'take Bharat into her care' and so save it—from abject poverty.

p. 242. *Krishan* : Hindu god.

p. 243. *the china can be washed* : Orthodox Hindus are extremely wary of contamination. They fear contamination from people of the lower castes, people of other religions, and from those who have a blemished character or reputation. Here Bhama feels that all the kitchen utensils and other objects handled by Saman need to be destroyed, since they have been contaminated by her touch. However her prudent housewifely instinct interferes with the religious impulse, and she orders that only the inexpensive earthenware be destroyed, while the china be reused after a thorough wash.

p. 244. *Sitaji...for years she had to face untold hardships in the forest* : Ramchandr was the seventh incarnation of the Hindu god Vishnu, for whom the Ramayana is named. He was the son of the ruler of Ayodhya, a north Indian state. He married Sita, the daughter of a Nepalese Raja. As the result of a conspiracy hatched by his stepmother, his father banished him from his home. Ram's faithful wife Sita accompanied him into the forests of South India, where they spent many years in great hardship. The king of Lanka (now Sri Lanka) abducted Sita, and carried her off to his island home. But she was rescued by her husband with the help of an army of monkeys. Following the death of his father, Ram returned to his home with Sita and was crowned king. But more intrigue followed and when it was discovered that Sita was pregnant, the intriguers convinced Ram that she had given in to the king of Lanka's sexual advances and was carrying his child. This time Sita was exiled by her husband. Once again she found herself in the wilderness, where she lived out the rest of her life in an ashram, or hermitage.

p. 244. *Savitri bore up in the face of tragedies* : In Hindu mythology, Savitri was the only child of a king. Since suitors were intimidated by her beauty and intelligence, she was unmarried. Finally her father asked her to find a husband for herself. Savitri left her home and after many years arrived in an ashram. In the ashram she met a young man who had arrived with his blind father. Her teacher told her that the young man was Satyavan and his blind father was a king who had lost his kingdom. Savitri decided to marry Satyavan, and went home to tell her father about her decision. But when the astrologer saw Satyavan's horoscope he said that Satyavan was destined to die one year from that date. Savitri's father was most perturbed and asked his daughter to choose a different man. However, Savitri told him that she had made up her mind and would happily share the one year with Satyavan, rather than marry someone else.

So the two were married and lived happily if frugally in the ashram, with his father. On the day that he was doomed to die, Satyavan went out into forest to collect firewood. Savitri insisted on accompanying him. Suddenly, Satyavan felt dizzy and lay down, with his head on Savitri's lap. As he closed his eyes, Savitri saw Yama, the god of death taking away his soul. Savitri followed the god, pleading with him to spare her husband. Yama was impressed by her love for her mate and said that though he could not fulfill this wish, he would grant her anything else that she asked for. Savitri immediately asked for the restoration of her father in law's kingdom and his eyesight. The wish was granted, but Savitri still followed Yama. Trying to persuade her to go back, Yama offered to fulfill one more wish. Savitri asked that her own father be granted many children, since so far she was his only child. This wish too was satisfied, but still Savitri continued her pursuit. Impatient now, the god told her she could have one more, last wish, and then she must go back. Savitri therefore made her last wish, 'Lord Yama, grant me many children by my husband Satyavan.'

Outwitted, the god had to let Satyavan go back to life, and the couple lived happily together for many years, as king and queen of Satyavan's kingdom. When they grew old, Yama came again and they went with him together, happily.

p. 249. *Katak* : Seventh month of the Hindu calendar, corresponding approximately to 15 October-15 November.